TRANSACTIONS

OF THE

AMERICAN PHILOSOPHICAL SOCIETY

HELD AT PHILADELPHIA
FOR PROMOTING USEFUL KNOWLEDGE

NEW SERIES—VOLUME 49, PART 1
1959

THE SPECTRUM OF BETA LYRAE

J. SAHADE, S.-S. HUANG, O. STRUVE, AND V. ZEBERGS

Berkeley Astronomical Department, University of California

THE AMERICAN PHILOSOPHICAL SOCIETY
INDEPENDENCE SQUARE
PHILADELPHIA 6

FEBRUARY 1959

Library of Congress Catalog
Card No. 59–7112

THE SPECTRUM OF BETA LYRAE [1]

J. Sahade,[2] S.-S. Huang, O. Struve, and V. Zebergs

CONTENTS

Beta Lyrae, known as an eclipsing variable since 1784, has been the subject of a large number of investigations, some of which were listed by Struve [1][3] in his 1957 Russell lecture. The problems posed by the remarkable spectroscopic behavior of this system have been discussed by Struve [1, 2, 3, 4] on a number of occasions, and the first concentrated effort to solve them was made at the Yerkes Observatory in 1940, and simultaneously by Z. Kopal at Harvard. These papers were published in Volumes 93 and 94 of the *Astrophysical Journal* [5, 6, 7, 8, 9, 10].

An attempt to gather more information on Beta Lyrae and to try to answer some of the remaining puzzles led to an observational effort by Struve and Sahade in the summer of 1955. The new material of 195 plates was obtained on Eastman Process emulsion with the 32-inch camera of the coudé spectrograph of the Mount Wilson 100-inch reflector which gives a dispersion of approximately 10 A/mm.

THE SPECTRUM

The spectrum of Beta Lyrae is composite and displays absorption as well as emission features. We distinguish:

(*a*) A set of absorption lines which corresponds to a star of spectral type B8. These lines remain reasonably constant in intensity and show a large variation in velocity throughout the cycle.

(*b*) A set of peculiar absorption lines, usually referred to as "the B2 or the B5 spectrum," and especially shown by He I and H, which displays several components and must originate in an expanding shell or ring which surrounds the entire system.

(*c*) Broad and red-displaced absorption lines of H, He I, Mg II, Si II, Ca II, and Fe II, which appear only before mid-eclipse, and are usually designated as the "red satellite lines." This set can be roughly described as resembling an A2-type star; it must originate in a stream of gas which moves from the secondary star toward the following hemisphere of the primary B8 component.

(*d*) Violet-displaced absorption lines which appear only after mid-eclipse, and are usually designated as the "violet satellite lines." These lines display some structure: in H, He I, Mg II, and Fe III they appear as a broad feature upon which a narrower line is approximately centrally superimposed, in Ni II the line appears only as a narrow feature, and in S II it is perhaps present only as a broad absorption. The ionization indicated by this set of lines is higher than that indicated by the red satellite lines. The origin of the violet satellite lines can be traced to a stream from the primary component toward the following hemisphere of the secondary star.

TABLE 1

WAVE LENGTHS USED IN THE DETERMINATION
OF THE RADIAL VELOCITIES

H	He I	C II	S III	Fe II
3679.355	3613.641	4267.15	3831.85	3748.489
82.810	3705.037	N II	Ca II	62.894
86.833	32.926	3838.39	3736.901	83.347
91.557	3819.637	3994.996	3933.664	3814.121
97.154	67.528	4176.164	68.470	3906.037
3703.855	71.819	Mg II	Ti II	35.942
11.973	88.646	4390.585	3685.192	4173.450
21.940	3926.530	4481.228	3741.633	78.855
34.370	64.727	Si II	57.684	4233.167
50.154	4009.270	3853.657	59.291	96.567
70.632	26.218	56.021	61.320	4303.166
97.900	4120.857	62.592	3900.546	51.764
3835.386	43.759	4128.053	4294.101	85.381
89.051	68.971	30.884	4300.052	4413.600
3970.074	4387.928	Si III		16.817
4101.737	4437.549	3796.11		89.185
4340.468	71.507	S II		91.401
		4153.098		4508.283
		62.698		15.337
				20.225
				22.634
				Fe III
				4419.59
				Ni II
				3769.455
				3849.58
				4067.051

[1] The observational material used in this investigation was obtained by O. Struve and J. Sahade as guest investigators at the Mount Wilson Observatory. The investigation was partly supported by grants from the Office of Naval Research and the National Science Foundation. The authors are indebted to the Computer Center of the University of California at Berkeley and its director, Dr. L. G. Henyey, for the use of the IBM 701 electronic computer.

[2] John Simon Guggenheim Fellow, February 1, 1955–February 1, 1957.

[3] Numbers in brackets indicate references on page 32.

1

TABLE 2
RADIAL VELOCITIES OF THE B8 COMPONENT

Date 1955	U.T.	Cycle and Phase (P)	Si II	H	He I	N II	Mg II	S II	S III	Ca II	Ti II	Fe II (all)	Fe II λ 4233	Fe III	Ni II	All Lines
April																
5	10h57m	0.0235	− 23.6	− 28.5	− 15.6	—	—	—	—	—	—	—	—	—	—	− 26.1
6	10 38	.0999	−115.8	− 98.3	−112.3	−114.1	−111.0	−112.1	—	−112.1	−109.2	−115.8	−118.3	−114.7	−118.2	−114.6
	12 16	.1051	−116.7	−101.2	−115.1	−121.1	−117.8	−121.4	—	−117.6	−111.1	−115.0	−115.5	−119.6	−114.9	−115.7
7	9 53	.1748	−169.3	−182.2	−164.2	−176.2	−174.9	—	—	−179.6	−166.5	−160.7	−172.4	—	—	−166.9
	10 55	.1781	−168.3	—	−164.8	−177.7	−177.7	—	—	−176.4	−152.0	−165.0	−174.0	—	—	−166.3
	11 56	.1814	−172.3	—	−166.4	−173.7	−170.8	—	—	−181.0	−168.7	−165.7	−171.1	—	−168.2	−169.2
8	9 51	.2520	−199.1	−198.6	−197.1	−204.3	−196.8	−210.9	—	−200.6	−165.4	−203.3	−197.7	—	−191.4	−200.1
	10 38	.2546	−201.4	−204.4	−197.0	−210.8	−198.0	−200.7	—	−201.3	−185.2	−197.0	−204.3	—	—	−199.1
	11 24	.2570	−201.1	−211.3	−200.6	−195.3	−196.6	—	—	−206.2	−193.6	−204.2	−196.4	—	−213.3	−203.5
	12 11	.2595	−199.1	−204.1	−198.7	−211.9	−202.6	—	—	−197.2	−192.3	−200.4	−204.4	—	—	−200.7
May																
5	10 00	2.3408	−188.0	−185.8	−177.9	−207.5	−189.1	−202.4	—	−189.7	−177.1	−194.3	−188.7	—	−205.2	−187.4
	10 49	.3434	−182.4	−183.7	−190.2	−186.4	−189.5	−189.3	—	−184.9	−169.1	−190.1	−177.9	−169.8	−189.2	−182.3
	11 42	.3462	−180.5	−180.4	−173.7	−168.0	−189.2	—	—	−187.7	−191.6	−186.2	−179.4	—	−196.3	−180.6
6	7 03	.4086	−142.7	−135.4	−128.3	−124.4	−141.1	—	—	−140.5	−144.5	−143.6	−150.4	—	−149.6	−139.1
	7 59	.4116	−137.9	−138.6	−129.0	−122.0	−135.9	—	—	−131.9	−122.3	−140.6	−143.1	—	−139.0	−135.9
	9 00	.4149	−134.0	−142.7	−127.1	−124.4	−135.4	—	—	−129.6	−135.7	−144.1	−144.0	—	−143.7	−137.1
9	7 25	.6418	+111.5	+111.3	+111.0	+110.7	+106.9	—	—	+109.9	+109.5	+115.6	+107.9	—	+111.3	+112.0
	8 27	.6451	+115.2	+117.5	+115.4	+121.9	+116.8	—	—	+113.8	+124.6	+128.9	+114.4	—	+129.3	+117.9
	9 28	.6484	+119.4	+120.3	+114.8	+117.4	+118.8	—	—	+117.1	+121.3	+122.6	+115.8	—	+128.4	+119.2
	10 30	.6517	+120.5	+119.5	+117.3	+112.1	+119.1	—	—	+118.8	+124.5	+124.6	+123.6	—	+123.6	+120.3
	11 36	.6553	+124.1	+126.8	+117.4	+117.0	+122.4	—	—	+123.0	+132.0	+128.3	+117.7	—	+141.5	+125.8
June																
3	5 21	4.5687	+ 37.8	+ 30.7	+ 31.7	+ 48.8	+ 41.3	—	—	+ 38.1	+ 43.6	+ 39.1	+ 33.3	—	+ 44.8	+ 36.1
	6 32	.5725	+ 43.9	+ 32.9	+ 39.3	+ 46.9	+ 36.0	—	—	+ 38.7	+ 48.6	+ 44.9	+ 32.5	—	+ 60.1	+ 42.0
	7 26	.5754	+ 45.5	+ 36.6	+ 39.8	—	+ 43.8	—	—	+ 47.2	+ 54.5	+ 51.7	+ 41.8	—	+ 45.3	+ 44.8
	8 40	.5794	+ 48.8	+ 39.4	+ 41.2	+ 46.7	+ 48.9	—	—	+ 48.6	+ 55.2	+ 55.8	+ 48.9	—	+ 54.5	+ 48.0
June																
3	9h43m	4.5828	+ 54.6	+ 42.0	+ 44.5	—	+ 49.8	—	—	+ 52.7	+ 61.6	+ 57.8	+ 54.6	—	+ 61.7	+ 50.9
	10 46	.5862	+ 56.3	+ 46.5	+ 47.1	+ 59.5	+ 52.5	—	—	+ 53.4	+ 60.9	+ 59.2	+ 53.8	—	+ 58.6	+ 53.4
	11 47	.5894	+ 56.6	+ 52.0	+ 53.2	—	+ 60.6	—	—	+ 57.5	+ 67.5	+ 62.5	+ 61.6	—	+ 65.3	+ 57.7
June																
4	6 30	.6498	+116.7	+126.0	+116.8	+116.9	+118.4	—	—	+116.0	+122.5	+123.3	+118.3	—	+155.1	+121.7
	7 34	.6532	+122.2	+127.5	+116.8	+121.2	+122.6	—	—	+121.9	+131.7	+128.1	+120.4	—	+128.6	+124.8
	8 34	.6564	+124.1	+128.1	+121.2	+119.7	+120.1	—	—	+123.3	+134.7	+128.3	+129.7	—	+125.5	+126.1
	9 32	.6595	+126.1	+132.1	+125.3	+126.8	+124.8	—	—	+121.5	+144.0	+130.9	+125.2	—	+129.9	+129.7
	10 25	.6624	+127.4	+132.7	+125.6	+121.2	+120.5	—	—	+126.5	+136.4	+132.5	+132.3	—	+125.3	+129.6
	11 18	.6652	+126.8	+129.6	+123.7	+121.4	+120.6	—	—	+130.7	+140.9	+135.8	+128.6	—	+123.0	+128.6
5	7 02	.7288	+159.4	+166.6	+163.7	+151.2	+153.3	—	—	+162.0	+180.3	+165.9	+159.2	—	+173.6	+164.5
	8 06	.7323	+158.9	+170.7	+162.9	+159.6	+154.3	—	—	+162.0	+176.9	+166.8	+162.0	—	+161.4	+165.7
	9 06	.7355	+160.9	+166.2	+162.4	+161.0	+155.2	—	—	+160.1	+171.6	+168.6	+163.3	—	+158.3	+164.7
	10 05	.7386	+163.9	+167.7	+166.6	—	+156.7	—	—	+158.2	+170.3	+175.7	+169.0	—	—	+167.0
6	11 11	.7422	+163.8	+169.5	+163.3	+157.0	+159.8	—	—	+160.5	+174.1	+173.7	+158.8	—	—	+167.0
	7 10	.8066	+164.9	+176.5	+167.7	—	+167.1	—	—	+160.0	+165.1	+162.6	+171.3	+171.4	+168.0	+168.2
	8 12	.8099	+160.6	+172.0	+168.5	+160.1	+154.5	—	—	+155.9	+186.7	+162.9	+157.4	—	—	+167.1
	9 10	.8130	+162.7	+168.5	+168.5	+170.0	+167.2	—	—	+158.7	+167.8	+169.2	+165.3	—	—	+167.6
	10 10	.8163	+159.3	+171.5	+166.9	+166.1	+159.4	—	—	+158.8	+170.8	+161.9	+165.2	—	—	+165.6
	11 12	.8196	+162.3	+166.3	+166.1	—	+161.1	—	—	+158.4	+150.5	+159.9	+155.0	—	—	+163.7
	11 58	.8210	+160.1	+168.3	+168.2	—	+155.7	—	—	+161.8	+153.6	+161.4	+161.4	—	—	+165.2
7	6 00	.8802	+130.4	+127.8	+139.4	+130.9	+131.4	—	—	+134.9	+142.0	+122.2	+106.2	+134.5	+128.8	+131.5
	6 46	.8826	+125.6	+130.4	+139.6	+130.9	+119.8	—	—	+122.1	+128.8	+124.6	+127.0	+137.9	+120.5	+130.9
	7 34	.8852	+126.5	+120.8	+136.2	+137.0	+121.7	—	—	+126.4	+117.3	+115.5	+118.9	—	—	+125.8
	8 29	.8882	+122.2	+126.8	+132.1	+126.9	+118.9	—	—	+116.6	+122.0	+116.8	+122.4	—	—	+124.2
	9 15	.8906	+120.4	+121.7	+131.8	+125.3	+118.1	—	—	+121.6	+121.9	+116.8	+111.5	—	—	+126.0
	9 46	.8923	+117.4	+135.1	+127.7	+126.1	+114.6	—	—	+115.0	+104.0	+114.3	+115.8	—	+107.4	+123.7
June																
7	10h18m	4.8940	+118.3	+125.1	+130.2	+118.3	+115.6	—	—	+113.8	+113.7	+111.9	+112.9	—	+109.6	+120.3
	10 51	.8958	+113.6	+124.2	+128.6	+122.9	+113.6	+110.9	—	+115.2	+113.7	+105.8	+115.8	—	+108.1	+118.8
	11 21	.8974	+113.5	+124.7	+123.0	+123.7	+110.8	—	—	+111.4	+108.7	+111.2	+104.8	—	+102.0	+117.7
	11 52	.8991	+114.9	+126.7	+131.8	—	+115.2	—	—	+119.4	+116.9	+110.7	+107.0	—	—	+123.2
8	5 17	.9552	+ 73.4	+ 70.0	+ 64.3	+ 68.0	+ 71.4	—	—	+ 72.9	+ 66.7	+ 67.7	+ 61.7	—	+ 66.2	+ 68.3
	6 21	.9586	+ 68.5	+ 54.0	+ 52.7	+ 65.6	+ 66.9	—	—	+ 65.1	+ 65.0	+ 62.9	+ 64.5	—	+ 66.6	+ 63.2
	7 12	.9614	+ 66.1	—	+ 59.3	—	+ 64.0	—	—	+ 69.6	+ 60.9	+ 68.9	+ 69.6	—	—	+ 65.2
	8 04	.9642	+ 63.3	+ 78.1	+ 50.6	—	+ 43.2	—	—	+ 62.2	+ 56.0	+ 51.9	+ 55.0	—	+ 52.4	+ 55.6
	9 05	.9675	+ 61.2	—	+ 52.4	—	+ 53.4	—	—	+ 61.3	+ 46.5	+ 51.7	+ 52.8	—	+ 29.8	+ 53.8
	10 15	.9712	+ 48.0	+ 30.1	+ 44.2	—	+ 54.8	—	—	+ 55.0	+ 50.9	+ 47.5	+ 40.4	—	+ 34.6	+ 45.3
	11 31	.9753	+ 35.5	+ 47.1	+ 50.4	—	+ 62.4	—	—	+ 49.0	+ 36.1	+ 37.3	+ 36.7	—	+ 39.3	+ 42.1
9	4 56	5.0314	− 25.9	− 42.3	− 33.1	—	− 28.7	—	−39.9	—	− 32.7	− 31.6	− 25.9	—	—	− 35.1
	6 31	.0365	− 32.5	− 44.6	− 26.1	—	− 34.3	—	—	− 29.0	− 28.5	− 35.2	− 33.2	—	− 38.7	− 36.7
	7 31	.0397	− 34.1	− 49.9	− 42.6	—	− 32.4	—	—	− 53.7	− 33.7	− 30.9	− 37.6	—	− 36.4	− 40.5
	8 34	.0431	− 37.1	—	− 42.6	—	− 39.3	—	—	− 37.3	− 46.0	− 32.4	− 41.4	—	− 28.4	− 43.6
	9 40	.0467	− 40.5	—	− 47.0	—	− 41.4	—	—	—	− 35.5	− 41.6	− 40.0	—	—	− 42.3
	10 32	.0495	− 51.6	—	—	—	− 40.2	—	—	—	− 50.3	− 47.2	− 46.6	—	—	− 48.7
	11 16	.0518	− 50.6	—	—	—	− 62.0	—	—	—	− 50.3	− 60.3	—	—	—	− 51.6
	11 52	.0538	− 56.0	− 57.6	− 57.4	—	− 51.2	—	—	− 56.3	− 38.8	− 55.5	− 56.8	—	—	− 55.4
10	5 04	.1092	−122.1	−105.3	−121.1	−123.2	−129.1	−134.8	—	−121.8	−115.3	−118.2	−116.5	—	−125.8	−116.7
	5 45	.1114	−123.1	−103.1	−119.5	−122.4	−124.6	−129.6	—	−118.4	−111.9	−120.3	−114.3	—	−123.5	−116.2
	6 21	.1133	−122.8	−103.2	−126.0	−124.2	−135.1	−137.1	—	−123.7	−119.0	−123.7	−120.8	—	−116.8	−123.7
	6 47	.1147	−126.1	− 99.9	−126.2	−127.1	−130.2	−136.4	—	−124.8	−121.6	−122.7	−126.0	—	−126.2	−126.3
	7 09	.1159	−126.5	− 98.3	−126.5	−127.1	−128.8	—	—	−125.5	−122.3	−122.8	−125.3	—	−124.0	−124.9
	7 30	.1170	−128.8	−106.4	−132.1	−133.3	−130.2	−145.9	—	−126.7	−121.4	−122.5	−123.1	—	−125.5	−128.5
	7 51	.1182	−130.7	−104.9	−127.4	−136.4	−135.7	—	—	−136.8	−129.9	−123.1	−136.9	—	−132.3	−133.9
	8 12	.1193	−129.0	−107.5	−130.9	−139.5	−133.7	−137.9	—	−131.4	−123.5	−123.1	−127.5	—	−125.6	−124.2
	8 33	.1204	−126.7	−112.6	−131.2	−126.5	−135.8	—	—	−130.2	−117.4	−127.0	−131.9	—	−137.6	−124.6
June																
10	8h54m	5.1215	−132.6	−109.7	−130.0	−134.2	−127.0	—	—	−130.8	−118.3	−129.0	−124.0	—	−130.2	−124.8
	9 25	.1232	−132.1	−102.8	−132.6	−128.9	−131.1	−141.0	—	−122.3	−114.6	−128.4	−132.0	−132.7	−131.6	−123.8
	9 56	.1249	−133.0	−110.6	−130.2	−129.0	−143.0	−143.3	—	−132.4	−122.4	−128.3	−128.5	−134.5	−123.5	−127.1
	10 27	.1265	−133.2	−111.6	−132.4	−132.8	−136.0	−141.8	—	−129.5	−123.9	−129.6	−132.1	−132.4	−131.0	−127.0
	10 59	.1282	−136.2	−112.8	−135.8	−127.5	−146.2	—	—	−133.5	−122.6	−130.0	−132.2	—	−131.1	−129.0
	11 30	.1299	−135.4	−114.7	−138.3	−146.7	−142.8	—	—	−136.4	−126.3	−132.2	−135.8	—	−122.9	−130.2
	11 58	.1314	−138.8	−110.8	−138.7	−135.2	−139.2	—	—	−130.2	−122.8	−131.4	−138.0	—	−131.9	−128.7

VOL. 49, PT. 1, 1959] THE SPECTRUM OF BETA LYRAE 3

TABLE 2—(*Continued*)

Date 1955	U.T.	Cycle and Phase (P)	Radial Velocities (Km./Sec.) from													
			Si II	H	He I	N II	Mg II	S II	S III	Ca II	Ti II	Fe II (all)	Fe II λ 4233	Fe III	Ni II	All Lines
July 4	4^h03^m	6.9621	+ 62.7	—	+ 63.9	+ 68.1	+ 67.8	—	—	+ 64.4	+ 60.4	+ 60.0	+ 59.3	—	+ 54.8	+ 62.7
	5 00	.9652	+ 57.1	—	+ 54.0	+ 61.9	+ 61.9	—	—	+ 55.2	+ 56.7	+ 55.3	+ 61.5	—	+ 61.6	+ 56.2
	6 16	.9693	+ 53.1	—	+ 44.7	—	+ 54.7	—	—	+ 62.8	+ 56.6	+ 50.4	+ 54.2	—	—	+ 54.4
	7 34	.9735	+ 46.9	—	+ 43.3	+ 54.7	+ 47.0	—	—	+ 49.3	+ 52.3	+ 49.4	+ 52.5	+ 35.8	—	+ 47.2
	8 25	.9762	+ 41.7	—	+ 42.0	—	+ 46.3	—	—	+ 51.1	+ 43.4	+ 44.6	+ 43.8	—	+ 45.5	+ 44.5
	9 13	.9788	+ 43.1	—	+ 43.4	—	+ 43.9	—	—	+ 53.6	+ 49.8	+ 42.0	+ 39.4	—	—	+ 44.3
	10 02	.9814	+ 39.4	—	+ 33.9	—	+ 37.9	—	—	+ 45.9	+ 36.7	+ 31.5	+ 33.5	+ 23.8	—	+ 36.3
	10 47	.9838	+ 32.7	+ 35.6	+ 27.3	—	+ 45.4	—	—	+ 43.2	+ 35.9	+ 27.8	+ 32.8	—	—	+ 33.5
	11 44	.9869	+ 28.6	—	+ 24.6	—	+ 33.7	—	—	+ 33.0	—	+ 32.1	+ 34.8	—	—	+ 29.8
5	3 55	7.0390	− 31.9	− 29.3	− 28.3	− 31.7	—	—	—	—	− 23.2	− 37.1	− 34.9	—	—	− 30.4
	4 49	.0419	− 34.8	− 35.9	− 29.9	—	− 39.7	—	—	—	− 25.0	− 36.2	− 41.5	—	—	− 33.9
	5 50	.0452	− 36.7	− 38.2	− 32.2	—	− 44.6	—	—	—	− 30.4	− 35.5	− 46.6	—	—	− 35.4
	6 41	.0480	− 31.7	− 24.8	− 37.7	− 45.7	− 45.3	—	—	—	—	− 37.9	− 45.2	—	− 51.6	− 35.8
	7 36	.0509	− 46.7	− 43.4	− 41.5	—	− 57.0	—	—	—	− 45.5	− 47.4	− 48.2	—	—	− 45.0
	8 37	.0542	− 59.5	− 56.9	− 53.3	—	− 61.9	—	—	—	—	− 65.3	− 58.4	—	—	− 58.2
	9 28	.0569	− 59.9	− 59.4	− 54.6	—	− 62.7	—	—	—	—	− 55.6	− 63.5	—	—	− 57.2
	10 27	.0601	− 60.5	− 57.1	− 62.3	—	− 58.7	—	—	—	− 50.8	− 61.9	− 58.6	—	—	− 59.3
6	3 52	.1162	−124.0	− 98.0	−115.9	−129.2	−132.5	−139.4	—	−121.5	−117.9	−122.4	−123.3	−128.4	−131.8	−115.9
	4 40	.1188	−124.5	−100.4	−115.7	−121.7	−130.8	−138.7	—	−121.8	−121.2	−122.6	−132.8	—	−118.4	−116.4
	5 21	.1210	−124.7	− 97.9	−116.0	−125.5	−131.8	−138.7	—	−119.6	−116.5	−123.4	−123.4	—	−131.5	−116.4
	6 02	.1232	−129.1	−101.0	−123.6	−120.2	−140.6	−132.2	—	−120.1	−108.7	−127.1	−132.9	−122.4	−129.4	−121.0
July 6	6^h43^m	7.1254	−130.3	−108.2	−121.8	−134.8	−141.7	−138.8	—	−127.1	−117.1	−128.5	−132.2	−123.8	−130.1	−124.9
	7 27	.1278	−132.1	−105.4	−124.4	−131.8	−140.0	−131.6	—	−129.0	−123.0	−127.9	−127.2	−131.5	−132.5	−123.0
	8 08	.1300	−133.1	−106.1	−125.3	−137.3	−134.5	−139.0	—	−132.9	−127.6	−128.4	−138.2	−133.0	−127.1	−124.9
	8 49	.1322	−137.6	−101.1	−128.5	−138.1	−138.7	−144.2	—	−135.4	−134.5	−129.9	−143.2	−131.6	−133.3	−126.9
	9 30	.1344	−135.4	−105.3	−127.5	—	−144.0	−139.1	—	−135.3	−134.7	−133.7	−139.0	−135.8	−134.9	−126.7
	10 16	.1368	−137.8	−109.3	−128.9	−140.6	−142.5	−137.0	—	−139.1	−132.7	−132.4	−142.0	−131.1	−131.3	−127.3
	11 15	.1400	−140.8	−111.4	−130.0	−139.0	−140.9	—	—	−131.7	−133.5	−139.1	−140.5	−129.0	−114.1	−127.6
	12 02	.1425	−143.4	−112.3	−128.5	—	−146.9	−137.1	—	−136.0	−134.9	−130.8	−141.4	—	—	−128.0
7	3 53	.1936	−174.8	−170.4	−176.8	−175.5	−184.4	—	—	−176.4	−171.8	−164.4	−175.4	—	−174.8	−173.0
	4 34	.1958	−176.3	−180.7	−174.6	−194.7	−181.4	−188.2	—	−191.1	−160.4	−170.4	−174.9	−149.5	−172.3	−175.5
	5 15	.1980	−179.6	−174.7	−176.2	−178.8	−183.5	−175.1	—	−187.2	−163.8	−177.4	−177.2	—	—	−176.8
	5 56	.2002	−178.6	−175.2	−180.6	−180.3	−182.6	−181.8	—	−186.0	−168.3	−180.8	−178.0	—	−193.5	−179.2
	6 35	.2023	−181.6	−172.9	−176.7	−178.0	−186.8	−188.4	—	−189.0	−166.2	−176.5	−186.7	—	−183.1	−176.9
	7 16	.2045	−183.4	−185.0	−183.9	−175.0	−185.1	—	—	−190.1	−175.8	−176.6	−180.3	—	−177.9	−181.4
	7 52	.2064	−183.3	−191.4	−184.1	−183.4	−183.0	—	—	−188.1	−175.7	−181.3	−186.8	—	−197.5	−184.3
	8 27	.2083	−179.7	−199.0	−185.5	−178.9	−187.6	—	—	−190.6	−171.8	−181.6	−188.3	—	—	−186.6
	9 00	.2101	−183.3	−182.8	−182.5	−195.0	−181.0	—	—	−195.3	−172.8	−180.8	−180.4	—	−177.2	−182.5
	9 35	.2120	−184.5	−193.8	−185.1	−181.3	−183.1	−181.3	—	−189.9	−173.6	−181.4	−184.1	—	—	−184.6
	10 11	.2139	−185.3	−194.0	−188.5	−190.5	−185.9	—	—	−194.0	−172.3	−182.6	−185.5	—	—	−187.8
	10 52	.2161	−185.3	−191.9	−186.1	—	−188.9	—	—	−185.4	−183.8	−190.6	−197.9	—	—	−188.3
	11 38	.2186	−189.1	−192.1	−186.7	—	−193.4	—	—	−188.0	−184.3	−194.9	−190.0	—	—	−190.1
8	3 48	.2707	−200.9	−207.1	−200.1	−211.5	−201.6	−206.8	—	−209.6	−207.2	−205.3	−202.0	—	−207.2	−204.4
	4 28	.2728	−203.8	−209.2	−196.8	−217.3	−193.5	−211.0	—	−207.9	−209.7	−203.4	−198.3	—	−211.0	−204.3
	5 02	.2747	−201.9	−215.0	−199.8	−215.7	−197.6	−201.1	—	−210.4	−185.1	−203.6	−206.4	—	−207.4	−204.6
	5 33	.2763	−202.3	−210.3	−198.6	−205.8	−204.4	—	—	−207.6	−201.9	−207.7	−215.1	—	−214.9	−204.6
	6 02	.2779	−202.7	−207.0	−199.1	−204.7	−198.4	−196.0	—	−209.9	−202.5	−206.9	−205.8	—	−211.9	−201.9
	6 35	.2797	−205.1	−212.8	−197.7	−195.9	−199.7	—	—	−205.8	—	−214.4	−218.8	—	−205.2	−202.8
	7 07	.2814	−200.0	−218.2	−198.1	−194.5	−193.7	—	—	−209.8	—	−214.2	−218.9	—	−194.0	−202.6
July 8	7^h39^m	7.2831	−202.3	−221.9	−201.9	−197.5	−196.5	—	—	−210.2	−205.4	−215.6	−203.7	—	—	−206.9
	8 10	.2848	−204.5	−211.8	−199.0	—	−199.2	—	—	−206.8	—	−199.6	−210.2	—	—	−202.3
	8 41	.2864	−204.2	−201.0	−201.1	−194.6	−196.5	—	—	−207.9	—	−197.9	−207.4	—	—	−201.3
	9 11	.2880	−202.8	−215.9	−202.1	−193.8	−205.4	—	—	−203.3	—	−213.3	−211.7	—	−220.4	−205.8
	9 44	.2898	−203.7	−209.4	−198.5	−206.6	−193.0	−205.1	—	−205.9	−203.6	−208.6	−210.4	—	—	−204.2
	10 16	.2915	−199.2	−205.0	−199.3	−214.4	−203.6	—	—	−204.4	−199.1	−211.8	−206.1	—	−231.5	−204.5
	10 51	.2934	−204.1	−209.2	−199.7	−208.9	−193.9	−198.6	—	−208.0	−204.1	−207.1	−209.1	—	−219.4	−204.9
	11 22	.2951	−200.9	−209.9	−195.0	−202.1	−204.3	—	—	−205.2	−204.5	−202.5	−200.4	—	—	−202.4
	11 53	.2967	−202.0	−207.9	−200.8	−205.7	−200.9	—	—	−203.1	−204.1	−204.6	−203.3	—	−212.9	−204.2
9	3 56	.3485	−183.2	−186.1	−180.3	−174.5	−174.8	−178.5	—	−191.5	−194.0	−189.2	−179.1	—	−187.2	−184.9
	4 41	.3509	−182.3	−182.0	−175.4	−191.4	−183.0	−180.0	—	−190.3	−185.8	−182.3	−187.1	—	−193.2	−181.7
	5 21	.3530	−181.6	−183.1	−174.2	−196.5	−177.9	—	—	−189.9	−182.7	−193.8	−188.6	—	−192.9	−183.7
	5 53	.3547	−177.7	−182.2	−174.9	−192.9	−185.5	—	—	−184.8	−181.2	−182.4	−175.6	—	−194.0	−180.9
	6 30	.3567	−179.6	−187.8	−173.2	−181.6	−175.0	−172.1	—	−182.9	−189.0	−188.8	−182.8	—	−185.9	−182.6
	7 02	.3585	−178.3	−187.6	−173.7	−174.7	−186.8	—	—	−187.4	−181.2	−189.7	−188.7	—	−190.4	−183.1
	7 29	.3599	−176.5	−181.6	−172.7	−164.8	−183.9	—	—	−179.7	−182.8	−189.2	−180.0	—	−194.9	−180.9
	8 00	.3616	−175.0	−175.8	−172.5	−155.7	−181.7	−173.6	—	−176.9	−191.8	−173.8	−191.0	—	−183.5	−177.6
	8 28	.3631	−175.0	−174.0	−166.8	−165.7	−168.7	—	—	−178.9	−177.7	−184.8	−185.2	—	−180.8	−173.7
	8 57	.3646	−175.8	−176.1	−169.3	−174.9	−176.5	—	—	−181.3	−177.4	−186.4	−178.8	—	−184.1	−176.6
	9 29	.3663	−173.3	−180.2	−169.0	−157.3	−175.4	—	—	−170.3	−172.4	−180.9	−187.5	—	−175.6	−175.5
	10 04	.3682	−175.4	−172.7	−164.7	−167.4	−173.9	−155.5	—	−176.7	−180.9	−184.9	−181.8	—	−179.9	−174.1
	10 36	.3699	−172.4	−176.9	−164.1	−163.6	−170.7	—	—	−173.4	−168.5	−181.2	−173.1	—	−173.4	−171.7
	11 08	.3717	−171.2	−175.3	−165.3	−163.7	−170.8	−167.3	—	−167.2	−175.5	−186.0	−173.2	—	−179.6	−173.7
	11 42	.3735	−168.2	−171.3	−160.8	−162.1	−166.0	—	—	−172.0	−165.7	−178.9	−171.1	—	−182.8	−170.0
10	3 48	.4254	−123.9	−129.4	−122.9	−120.4	−133.9	—	—	−126.1	−137.6	−133.3	−130.2	—	−142.2	−129.5
	4 28	.4275	−121.3	−123.4	−121.2	−136.2	−122.9	—	—	−126.6	−124.0	−126.2	−130.5	—	−140.3	−124.4
	5 00	.4292	−122.0	−122.7	−120.5	−139.0	−114.0	—	—	−128.1	−120.3	−122.6	−131.7	—	−131.9	−122.5
July 10	5^h29^m	7.4308	−120.2	−122.7	−120.4	−109.1	−127.2	—	—	−126.2	−116.3	−122.8	−123.2	—	−139.1	−122.2
	5 58	.4324	−116.9	−122.2	−118.9	−101.4	−119.6	—	—	−121.8	−120.9	−125.8	−124.6	—	−127.2	−121.4
	6 30	.4341	−114.7	−120.5	−114.5	−110.6	−116.2	—	—	−116.5	−120.9	−123.2	−123.2	—	−132.7	−119.0
	7 00	.4357	−110.0	−120.6	−117.7	−124.0	−114.7	—	—	−116.0	−121.9	−120.3	−129.8	—	−131.2	−119.4
	7 31	.4373	−110.3	−116.2	−113.5	−103.8	−112.5	—	—	−115.6	−121.2	−120.8	−121.5	—	−128.5	−116.7
	8 05	.4392	−110.7	−108.8	−104.2	−107.0	−116.4	− 99.7	—	−110.5	−119.8	−121.1	−120.5	—	−117.7	−114.2
	8 35	.4408	−106.0	−116.1	−100.1	−108.5	−105.6	—	—	−115.9	−116.4	−115.1	−109.6	—	−125.6	−111.1
	9 06	.4425	−108.2	−108.0	−104.4	− 91.0	−110.5	—	—	−117.7	−116.1	−113.4	−115.1	—	−112.4	−110.1
	9 35	.4440	−106.7	−109.0	− 99.4	−108.4	−107.2	—	—	−106.3	−111.0	−111.3	−107.6	—	−112.2	−107.5
	10 10	.4459	−104.5	−107.5	−101.1	− 94.2	−109.0	—	—	−108.5	−109.2	−116.1	−106.9	—	−114.7	−109.4
	10 46	.4478	−103.8	− 98.0	− 95.6	− 91.1	−109.7	—	—	−108.4	−106.7	−108.2	−113.4	—	−105.2	−103.8
	11 22	.4500	− 94.7	−105.8	− 89.1	−112.2	− 98.0	—	—	−103.9	−100.9	−111.6	−102.7	—	− 93.6	−101.9
	11 56	.4516	− 95.3	−103.2	− 95.2	− 95.0	—	—	—	−102.2	−109.0	−104.2	−108.5	—	−102.7	−100.8
11	3 56	.5031	− 37.3	− 37.9	− 38.0	− 45.1	− 44.9	—	—	− 39.0	− 39.0	− 40.7	− 41.6	—	− 41.2	− 39.5
	4 50	.5060	− 32.8	− 33.9	− 31.2	− 38.2	− 35.0	—	—	− 35.0	− 32.9	− 38.6	− 34.5	—	− 32.4	− 34.7
	5 38	.5086	− 30.0	− 30.0	− 31.9	− 36.8	− 33.7	—	—	− 28.5	− 27.8	− 34.0	− 30.2	—	− 30.9	− 31.6

TABLE 2—(Continued)

Date 1955	U.T.	Cycle and Phase (P)	Radial Velocities (Km./Sec.) from													
			Si II	H	He I	N II	Mg II	S II	S III	Ca II	Ti II	Fe II (all)	Fe II λ 4233	Fe III	Ni II	All Lines
Sept. 7	6h18m	.5108	− 26.9	− 32.6	− 32.1	− 22.9	− 31.0	− 44.0	—	—	− 28.8	− 32.6	− 33.1	− 35.2	− 28.5	− 31.5
	7 00	.5130	− 25.8	− 28.6	− 24.9	− 28.3	− 27.6	—	—	− 27.8	− 23.3	− 29.9	− 25.3	—	− 24.4	− 27.3
	7 38	.5151	− 23.0	− 25.7	− 24.1	− 27.6	− 21.8	—	—	− 22.1	− 24.5	− 25.6	− 28.2	—	− 26.5	− 25.0
	8 17	.5172	− 19.6	− 21.1	− 23.0	− 18.9	− 30.8	—	—	− 18.9	− 18.6	− 22.1	− 26.1	—	− 23.8	− 21.4
	8 51	.5190	− 20.0	− 23.7	− 22.6	− 24.8	− 22.2	—	—	− 20.6	− 21.0	− 22.1	− 17.4	—	− 23.0	− 22.3
	9 29	.5210	− 17.3	− 19.6	− 17.2	− 28.9	− 16.4	—	—	− 16.6	− 18.8	− 19.6	− 17.5	—	− 12.4	− 18.8
	10 14	.5234	− 15.8	− 19.1	− 17.3	− 13.2	− 13.3	—	—	− 26.5	− 15.1	− 14.5	− 13.3	—	− 11.6	− 16.2
	10 52	.5255	− 12.3	− 14.2	− 16.1	—	− 15.3	—	—	− 11.8	− 9.0	− 13.4	− 18.4	—	− 11.5	− 13.0
	11 28	.5274	− 10.5	− 19.8	− 16.0	− 13.3	− 15.2	—	—	− 3.7	− 10.2	− 9.9	− 9.1	—	− 7.0	− 13.3
	12 00	.5291	− 7.1	− 16.7	− 8.5	− 18.8	− 9.3	—	—	− 4.5	− 5.0	− 8.0	− 4.7	—	− 3.2	− 10.9
	3h22m	11.9872	+ 30.7	+ 10.9	+ 22.6	—	+ 42.3	—	—	+ 36.1	+ 29.0	+ 24.2	+ 26.8	—	+ 5.1	+ 21.9
	4 43	.9915	+ 25.1	+ 8.2	+ 25.6	+ 29.0	+ 31.2	—	—	+ 36.2	+ 28.4	+ 26.6	+ 20.8	—	+ 6.4	+ 22.4
	6 15	.9965	+ 14.9	+ 3.0	+ 11.3	—	+ 27.0	—	—	+ 12.5	+ 4.8	+ 8.3	+ 9.1	—	− 0.5	+ 6.7
	8 00	12.0021	+ 11.4	+ 1.7	+ 12.6	—	+ 18.1	—	—	+ 25.2	+ 2.1	+ 13.0	+ 17.7	—	—	+ 8.4

(e) Broad emissions, especially in H and He I, which give the impression of being formed by a broad feature upon which a stronger, narrower emission is superimposed. The narrower feature stands out very clearly during the eclipse.

There are striking changes with phase and cycle in the spectrum. Struve [11] has already described some of them, but the great complexity of the spectrum of Beta Lyrae makes it impossible to give a description which would do justice to the wealth of information contained in the material. Therefore, an atlas is appended at the end of this paper containing photographic enlargements of our spectrograms in the region λλ3680–4580. An effort has been made to line up the strips, referring them to the lines of the B8 star. Mrs. Zebergs is responsible for the photographic work involved in the preparation of the atlas. The phases indicated at the margins are in fractions of the period and were computed by means of Prager's formula [12] as modified by Saidov [13], namely,

$$\text{Min} = \text{JD } 2398590.57 + 12.908006\,E + 0.3919 \times 10^{-5}E^2 - 0.3 \times 10^{-10}E^3.$$

The number preceding the decimal point stands for the cycle which was arbitrarily assumed to be zero for our first spectrogram, taken on April 5, 1955, at 10^h57^m U.T., and corresponding to phase 0.0235 P (this spectrogram is not reproduced in the atlas). Table 2 gives the correspondence between calendar dates and U.T. of mid-exposure and the phases.

THE RADIAL VELOCITIES

THE B8 SPECTRUM

All our spectrograms have been measured for radial velocity by making settings on the absorption lines, the laboratory wave lengths and identifications of which are listed in table 1. The results are presented in table 2 and are plotted in figures 1–12.

The radial velocities of Si II give the most consistent set of values, with extremely small scatter except during the eclipse. Furthermore, they seem to be free from any blending effects. Therefore, they were adopted for the determination of the orbital elements, as described in a later section of this paper. The velocity curve predicted from these orbital elements (listed in table 9, last column) is shown as a solid line in figures 1–12.

The velocities from the rest of the elements, except perhaps N II and Mg II, and even from the mean of all elements, depart slightly from the velocity curve as derived from Si II, and affect mainly the amplitude. It is probable that these departures are the result of the streams which are present in the system. The behavior of the radial velocities, especially of H and He I, in the region of, and immediately after maximum positive velocities, resembles the effect of the gaseous streams in AU Monocerotis [14] and U Coronae Borealis [15].

In several phase intervals the velocities from H and He I show a systematic trend. For instance, immediately after secondary minimum the lines of the Balmer series yield progressively more negative values toward the lower members of the series; between primary minimum and first quadrature both the H and He I lines give progressively more positive velocities toward the red part of the spectrum. Even Si II shows the effect of the streams at and immediately after primary minimum in the sense that the redder the wave length the more positive is the velocity, but there are no indications of similar effects outside primary eclipse.

The departures of the mean radial velocities of H and He I immediately after secondary minimum are worthy of further consideration. These means fall below the velocity curve from Si II and from other elements, and suggest a motion away from the B8 star, perhaps indicating the existence of gases streaming out from the system through the external Lagrangian point which is in front of the primary B8 component of the system. This interpretation agrees with the spectroscopic behavior at the relevant phases.

The departure of the velocities of Fe II from the velocity curve of Si II is probably spurious: no such

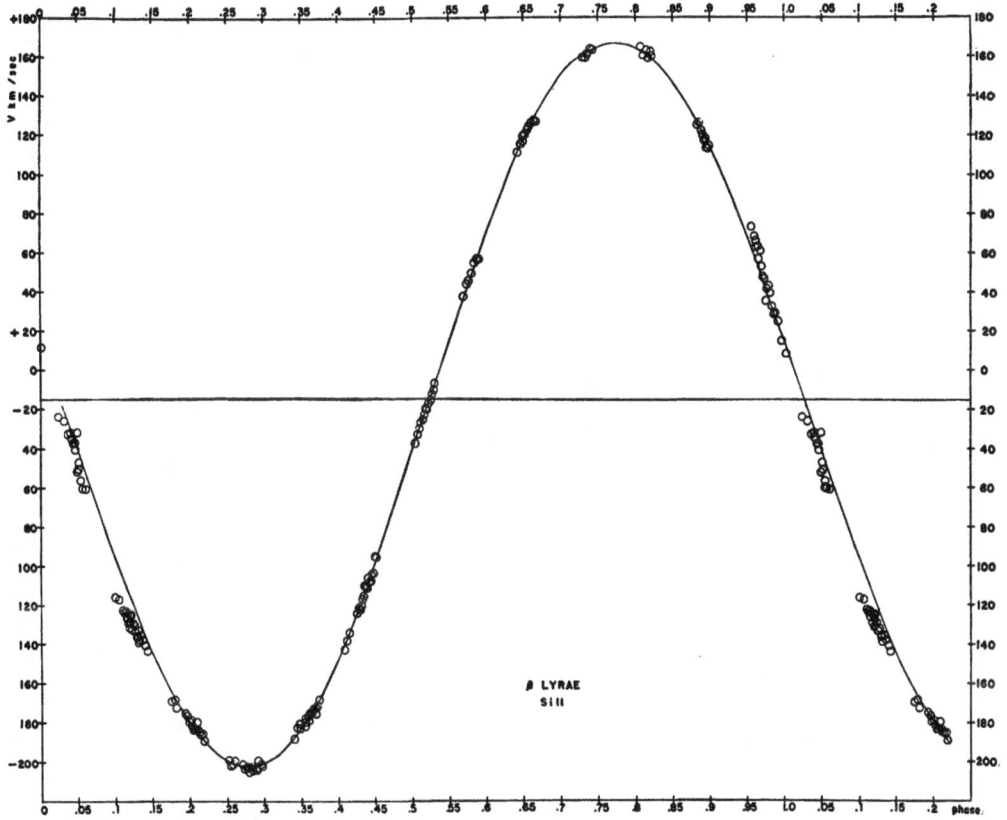

FIG.1. Radial velocities and velocity curve from Si II.

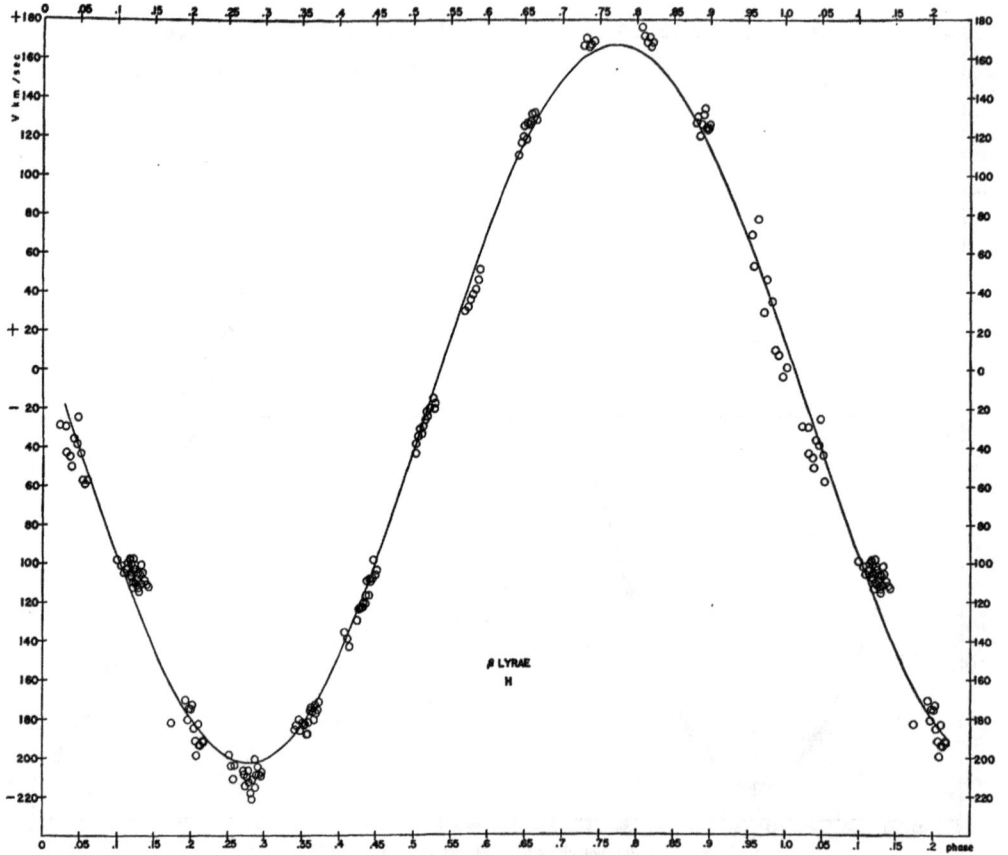

FIG. 2.   Radial velocities from H and velocity curve from Si II.

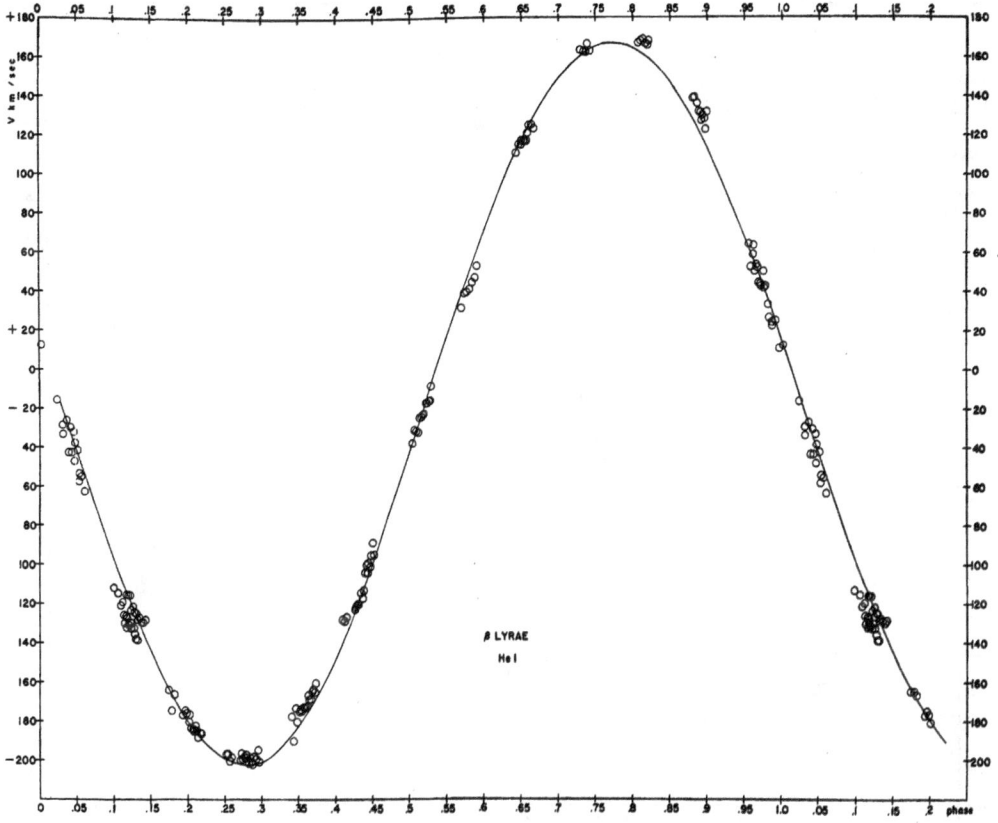

FIG. 3. Radial velocities from He I and velocity curve from Si II.

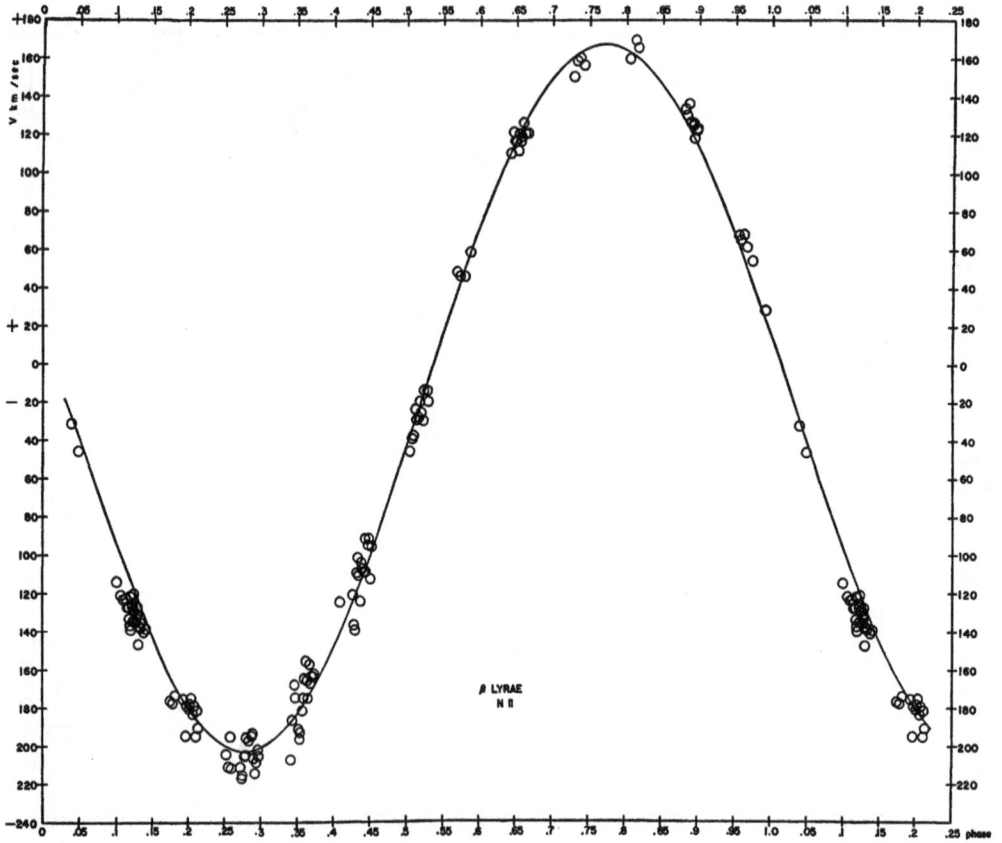

FIG. 4.   Radial velocities from N II and velocity curve from Si II.

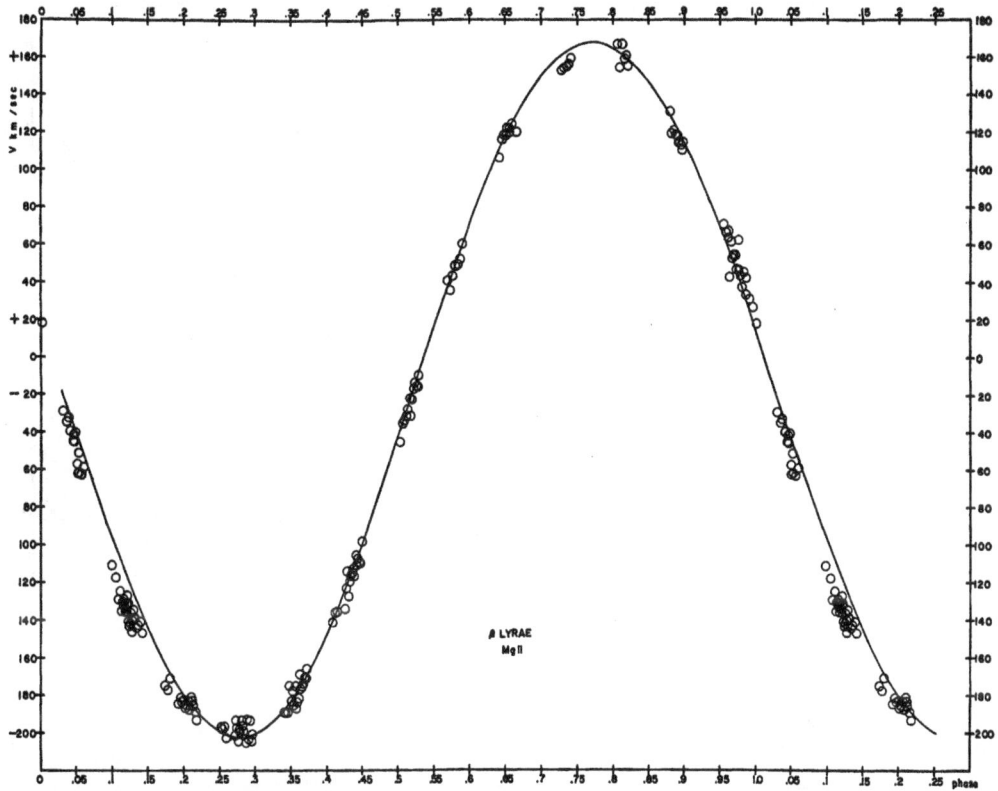

FIG. 5.   Radial velocities from Mg II and velocity curve from Si II.

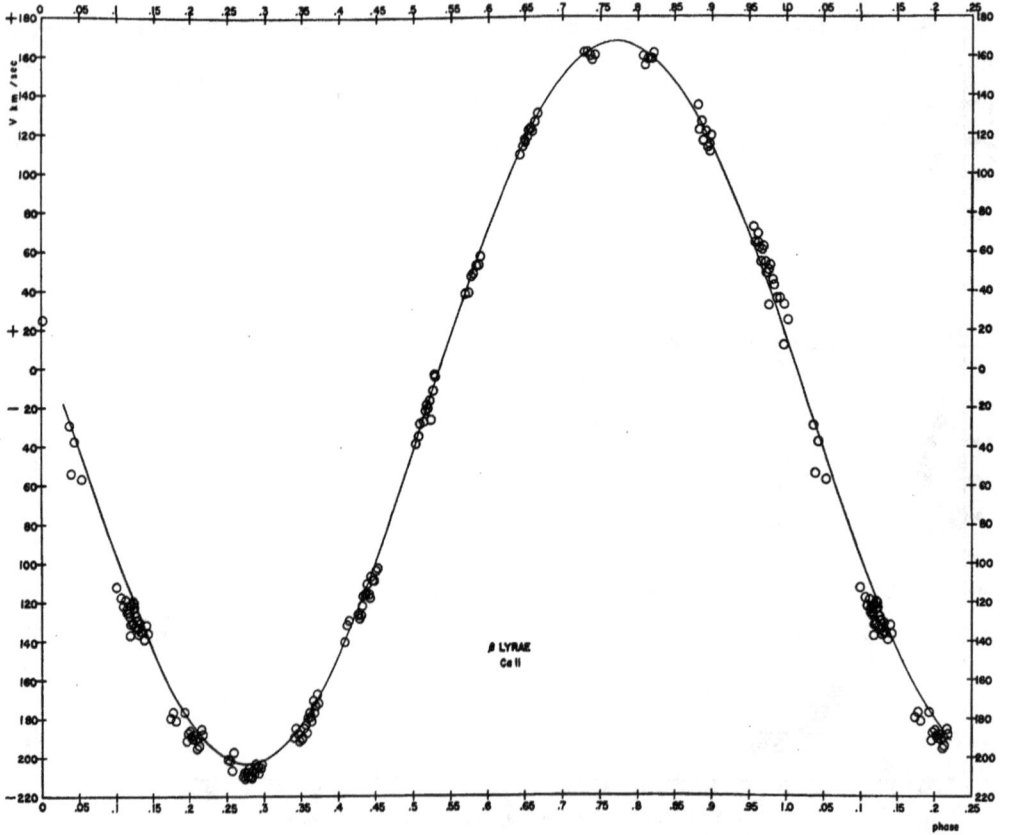

FIG. 6.   Radial velocities from Ca II and velocity curve from Si II.

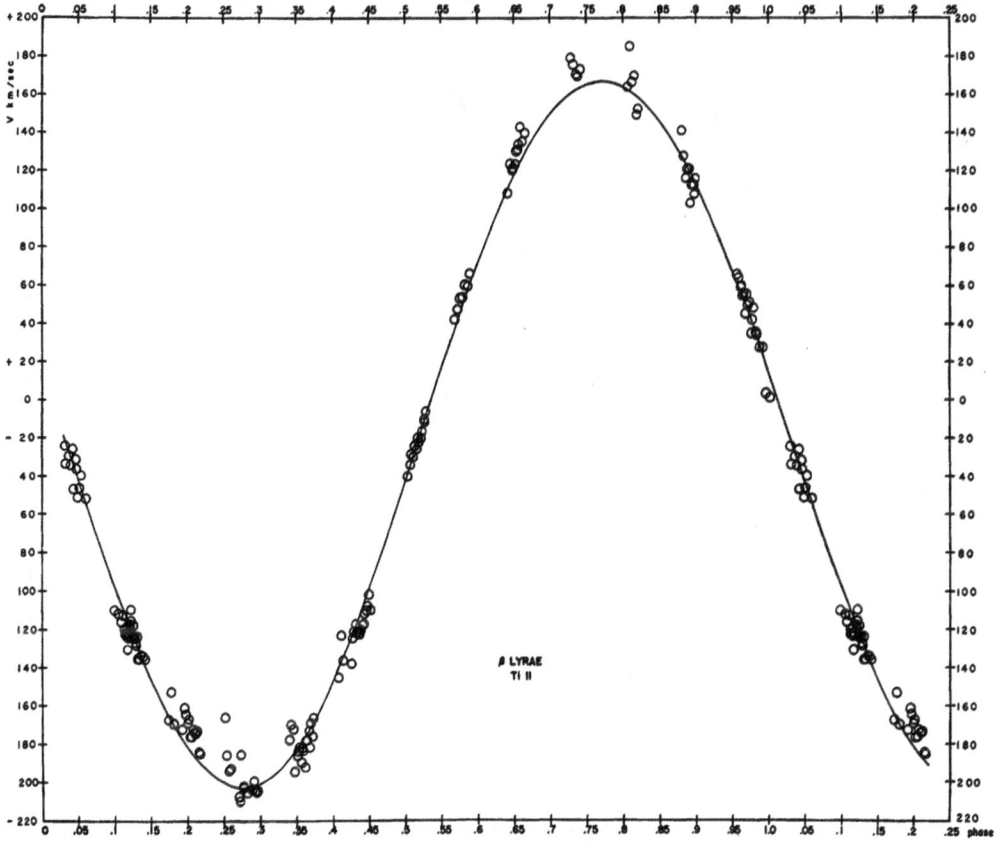

FIG. 7.　Radial velocities from Ti II and velocity curve from Si II.

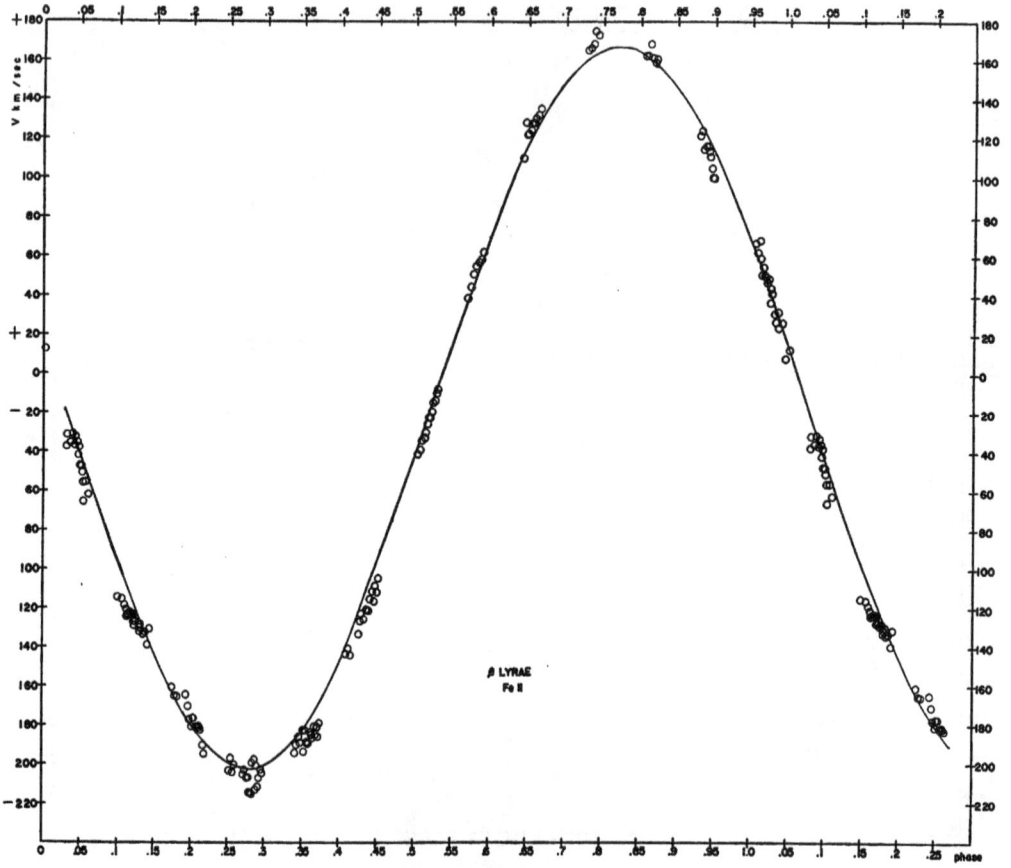

FIG. 8.   Radial velocities from Fe II and velocity curve from Si II.

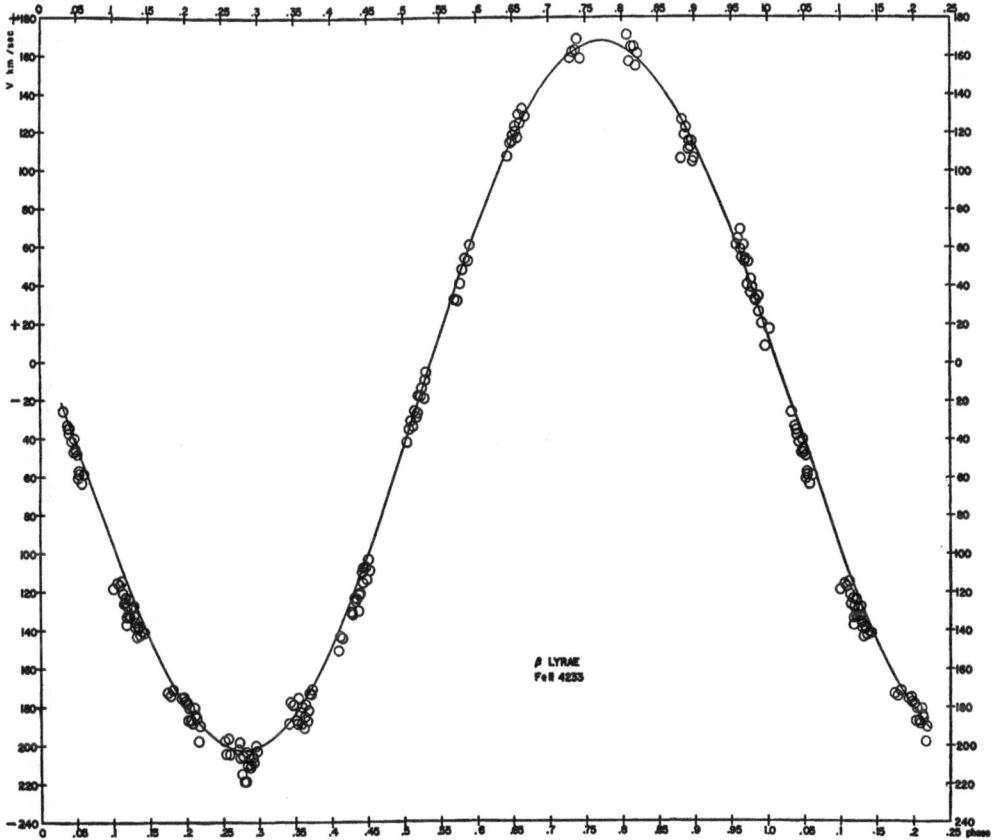

FIG. 9.   Radial velocities from Fe II λ4233 and velocity curve from Si II.

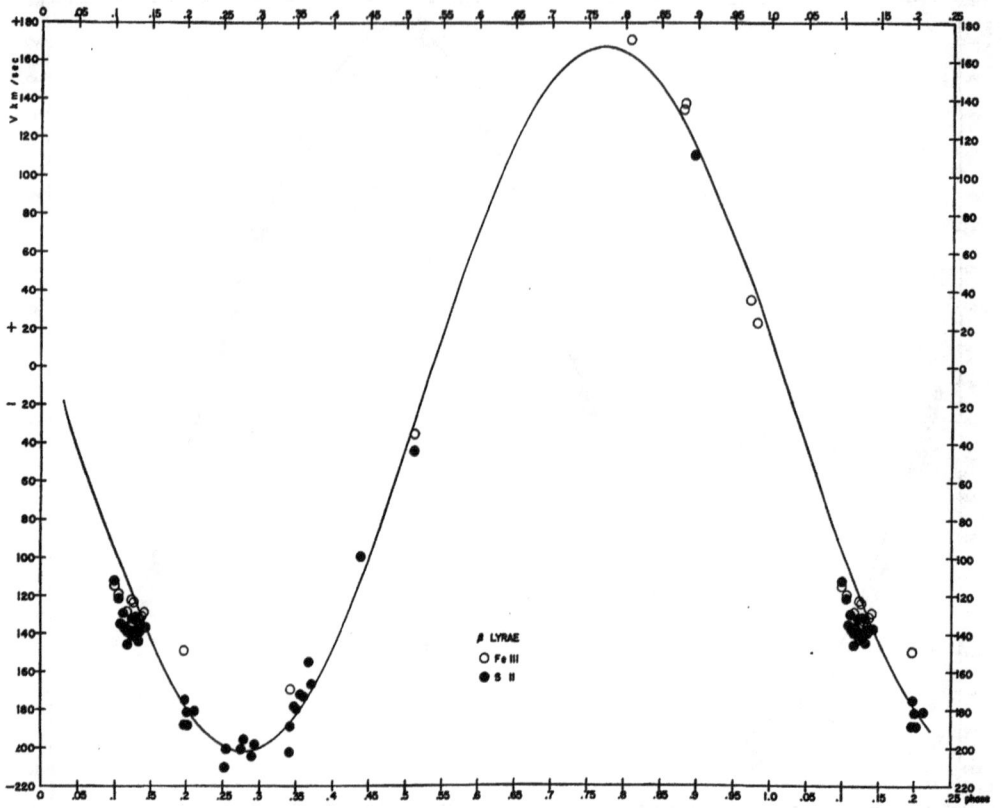

FIG. 10. Radial velocities from S II and Fe III and velocity curve from Si II.

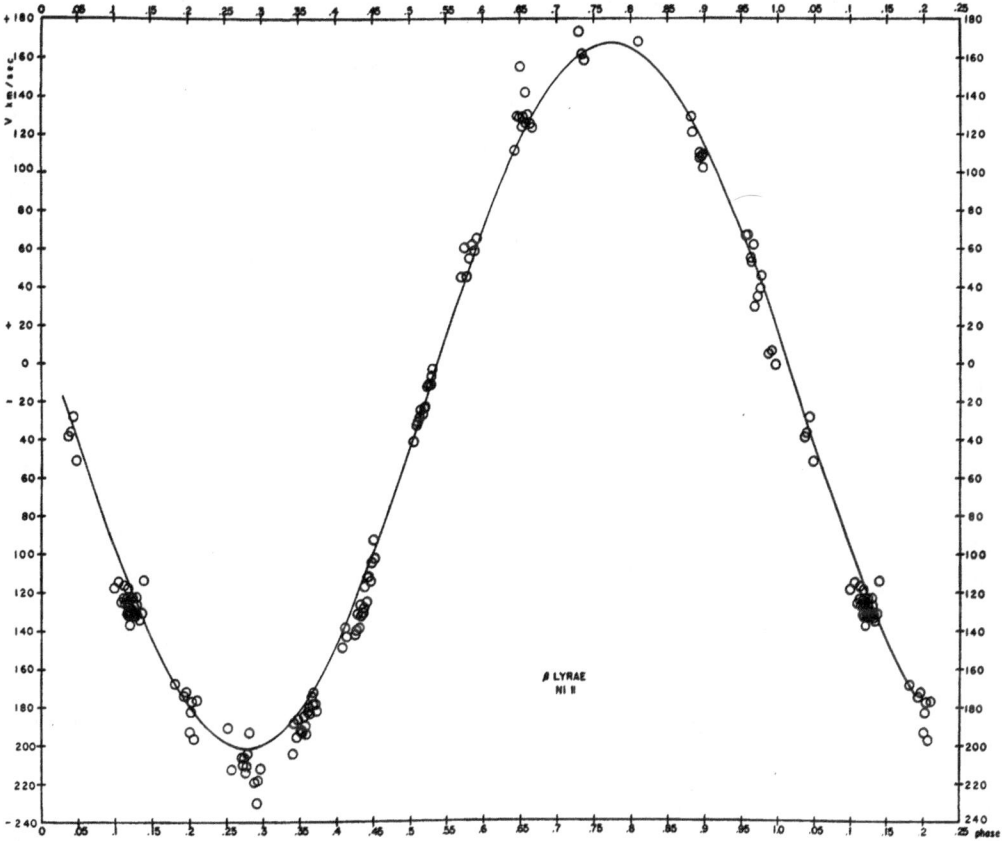

FIG. 11. Radial velocities from Ni II and velocity curve from Si II.

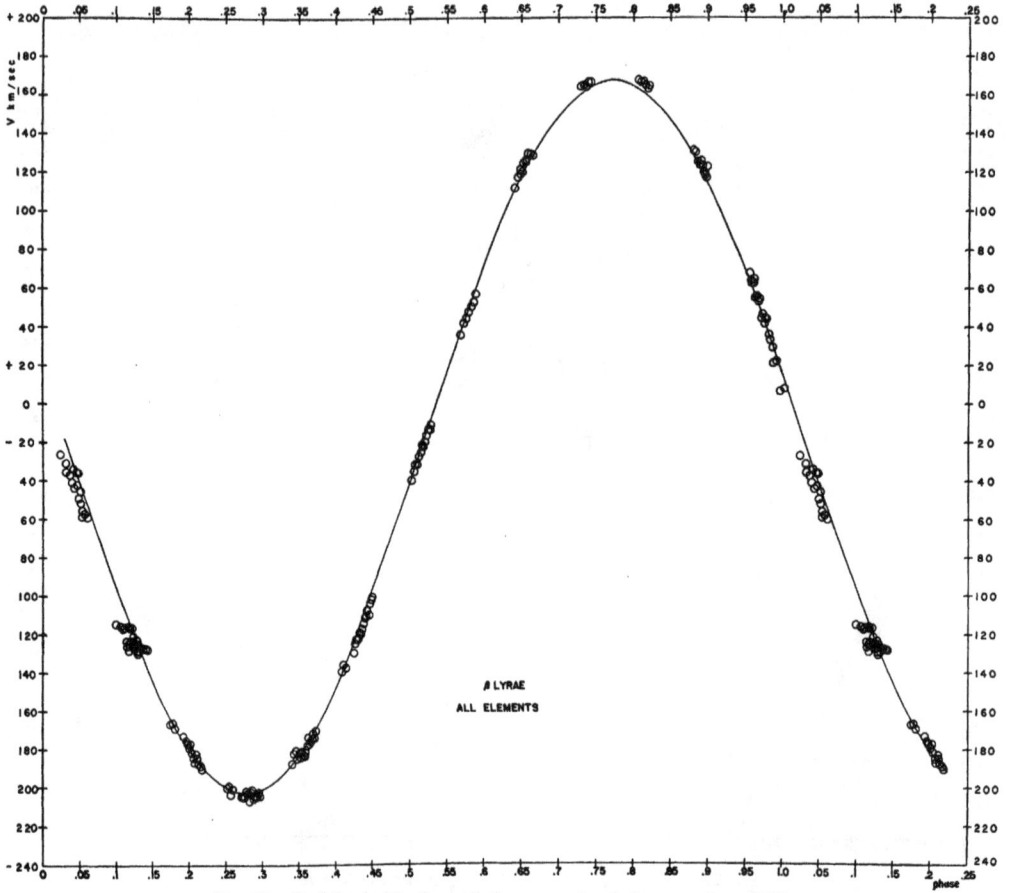

FIG. 12.   Radial velocities from all elements and velocity curve from Si II.

TABLE 3

RADIAL VELOCITIES FROM SHELL LINES OF HE I

Date 1955	U.T.	Cycle and Phase (P)	Radial Velocities (Km./Sec.) from He I λ3888	He I λλ3614 and 3965
Apr. 5	10h57m	0.0235	−179.2; −119.3	−128.3
	10 38	.0999	−172.0; −132.7	−125.9
	12 16	.1051	−164.9; −134.2	−128.2
7	9 53	.1748	−160.9; −138.8; −104.2	−139.7
	10 55	.1781	−163.3; −127.4; −96.5	−134.4
	11 56	.1814	−157.0; −132.7; −103.5	−140.6
8	9 51	.2520	−163.2; −121.5; −72.8	−135.1
	10 38	.2546	−164.1; −127.1; −76.0; +12.1	−133.6
	11 24	.2570	−159.4; −80.7; +18.4	−123.6
	12 11	.2595	−154.8; −84.8	−132.2
May 5	10 00	2.3408	−172.1; −139.1; −99.0; −57.3	−136.9; −100.6
	10 49	.3434	−169.9; −138.4; −98.3; −52.6	−146.9; −100.7
	11 42	.3462	−170.0; −136.0; −98.3; −53.4	−156.3; −100.7
6	7 03	.4086	−169.8; −140.6; −102.9; −38.4	−146.9; −103.7; −12.7
	7 59	.4116	−168.2; −102.1; −58.1; −23.4	−139.2; −105.3
	9 00	.4149	−168.3; −99.1; −32.2	−143.2; −101.3
9	7 25	.6418	−152.1; −92.3	−135.7; −92.5
	8 27	.6451	−171.7; −139.6; −97.8	−141.9; −103.3
	9 28	.6484	−169.5; −138.0; −103.4	−145.1; −103.4
	10 30	.6517	−169.6; −133.4; −101.9	−139.0; −99.7
	11 36	.6553	−170.5; −142.1; −98.9; −61.1	−136.0; −101.3
June 3	5 21	4.5687	−171.5; −137.7; −83.4	−135.6; −85.5
	6 32	.5725	−167.7; −137.0; −85.1	−138.0; −81.7
	7 26	.5754	−166.2; −137.9; −91.5	−137.4; −86.4
	8 40	.5794	−152.9; −88.4	−134.4; −75.7
	9 43	.5828	−151.5; −90.9	−139.1; −78.9
	10 46	.5862	−165.7; −139.8; −88.6	−138.4; −83.7
	11 47	.5894	−172.9; −142.2; −92.7	−141.6; −76.0
4	6 30	.6498	−156.9; −104.9	−145.2
	7 34	.6532	−152.2; −86.2	−144.5; −99.0
	8 34	.6564	−146.0; −97.3; −69.0; +13.6	−142.5; −83.7
	9 32	.6595	−150.1; −82.4; +15.1	−142.4; −73.0
	10 25	.6624	−150.2; −100.6; −67.6; +7.9	−143.3; −103.1; −64.6
5	11 18	.6652	−151.1; −86.6; +15.0	−142.6; −84.7
	7 02	.7288	−151.7; −84.8; −21.9	−132.4; −74.5
	8 06	.7323	−152.5; −78.6	−138.6; −82.3
	9 06	.7355	−151.9; −88.1	−141.0; −82.4
	10 05	.7386	−149.6; −87.5	−141.1; −85.6
	11 11	.7422	−150.5; −75.0	−143.6; −80.3
6	7 10	.8066	−146.4; −89.7	−136.4; −89.4
	8 12	.8099	−147.5	−124.9
	9 10	.8130	−144.2	−119.6
	10 10	.8163	−136.4	−119.7
	11 12	.8196	−143.6	−123.7
	11 58	.8210	−146.8	−127.7
7	6 00	.8802	−166.2; −127.7	−122.0
	6 46	.8826	−167.8; −123.7	−125.0
	7 34	.8852	−169.5; −130.9; −52.3	−127.5
	8 29	.8882	−168.8; −129.4	−125.2
	9 15	.8906	−169.7; −128.6	−124.6
	9 46	.8923	−172.0; −124.8	−128.4
	10 18	.8940	−173.7; −129.6	−137.0
	10 51	.8958	−173.7; −128.9	−135.5
	11 21	.8974	−175.4; −131.3	−133.3
	11 52	.8991	−171.4; −129.7	−127.9
8	5 17	.9552	−167.2; −127.9	−145.3
	6 21	.9586	−172.8; −131.1	−143.9
	7 12	.9614	−171.2; −131.9	−143.9
	8 04	.9642	−172.9; −137.5	−148.6; −50.6
	9 05	.9675	−175.4; −130.5	−151.0
	10 15	.9712	−177.0; −135.3	−148.8; −55.4
	11 31	.9753	−170.8; −139.4	−149.7
9	4 56	5.0314	−170.5; −134.4	−156.3
	6 31	.0365	−167.5; −128.2	−160.3
	7 31	.0397	−163.7; −131.4	−168.1
	8 34	.0431	−173.2; −131.5	−161.2; −64.0
	9 40	.0467	−163.9	−171.4; −83.4
	10 32	.0495	−173.4; −131.7	−173.8; −82.0
	11 16	.0518	−174.2; −132.5	−166.1; −84.3
	11 52	.0538	−174.3; −136.5	−172.3; −112.2
10	5 04	.1092	−176.4; −140.2	−170.5; −118.8
	5 45	.1114	−174.0; −138.6	−161.2; −112.6
	6 21	.1133	−171.7; −134.8; −58.5	−162.9; −108.9
	6 47	.1147	−178.8; −137.9; −58.5	−164.4; −107.3
	7 09	.1159	−184.3; −128.5; −56.1	−164.4; −114.3
	7 30	.1170	−187.5; −136.3; −53.0	−160.6; −112.0
	7 51	.1182	−181.3; −132.5; −68.8	−163.0; −109.0
	8 12	.1193	−186.0; −133.3; −57.0	−159.9; −104.3
	8 33	.1204	−178.1; −136.4; −62.5	−167.6; −115.1
	8 54	.1215	−183.7; −146.0; −63.4	−164.6; −113.7
	9 25	.1232	−178.2; −131.8; −51.6	−166.2; −115.2
	9 56	.1249	−183.0; −129.6; −73.7	−162.4; −113.8
	10 27	.1265	−179.1; −130.3; −60.3	−160.9; −117.7
	10 59	.1282	−177.7; −138.3; −80.1	−157.9; −113.2
	11 30	.1299	−182.4; −145.4; −84.8	−164.8; −108.5
	11 58	.1314	−180.0; −136.7; −71.4	−161.0; −108.5
July 4	4h03m	6.9621	−170.9; −108.7; −11.2	−169.8; −109.7; −10.1
	5 00	.9652	−174.0; −108.7; −26.1	−166.0; −112.0; +13.0
	6 16	.9693	−170.2; −106.5; −26.2	−173.0; −110.5; −8.7
	7 34	.9735	−170.4; −111.4; −17.0	−169.4; −113.8; −26.6; +29.7
	8 25	.9762	−173.5; −106.7; −23.3	−170.1; −114.6; +11.9
	9 13	.9788	−171.3; −109.9; −20.2	−172.6; −112.4; +4.1
	10 02	.9814	−175.3; −110.0; −25.0	−175.7; −118.7; −12.2
	10 47	.9838	−173.7; −108.4; −25.0	−168.8; −115.6; +0.2
5	11 44	.9869	−172.2; −109.3; −29.1	−169.7; −111.0; −15.4
	3 55'	7.0390	−175.8; −113.6; −14.5	−169.3; −113.7
	4 49	.0419	−169.6; −116.9; −19.3	−164.0; −114.6; −11.2
	5 50	.0452	−171.3; −118.8; −19.4	−169.5; −103.1
	6 41	.0480	−172.0; −116.2; −15.5	−161.8; −113.2
	7 36	.0509	−170.6; −117.9	−164.9; −114.0
	8 37	.0542	−171.5; −118.0	−169.7; −118.8
	9 28	.0569	−170.0; −112.6; −14.2	−116.5
	10 27	.0601	−174.0; −121.3; −26.9	−162.9; −113.6; −17.9
6	3 52	.1162	−170.6; −119.4	−117.9
	4 40	.1188	−172.2; −115.6	−118.8; −64.7
	5 21	.1210	−167.5; −114.8	−117.2
	6 02	.1232	−171.6; −119.6	−119.6
	6 43	.1254	−173.1; −114.9	−119.6
	7 27	.1278	−168.5; −117.4; −32.4	−119.0
	8 08	.1300	−171.8; −118.3; −32.5	−120.6
	8 49	.1322	−173.3; −116.7; −20.7	−122.1
	9 30	.1344	−171.9; −119.1; −16.1	−121.5
	10 16	.1368	−169.6; −121.6; −44.5	−120.8; −77.6
	11 15	.1400	−169.6; −124.0; −57.9	−119.3
	12 02	.1425	−169.7; −121.7	−122.4
7	3 53	.1936	−173.2; −121.3; −27.7	−120.5
	4 34	.1958	−171.7; −125.2; −12.0	−126.7
	5 15	.1980	−171.8; −123.0; −17.6	−121.4
	5 56	.2002	−173.4; −122.3; −16.1	−126.9
	6 35	.2023	−173.4; −121.5; −7.5	−126.1
	7 16	.2045	−172.0; −125.5; −23.3	−120.6
	7 52	.2064	−172.0; −121.6; −5.2	−124.7
	8 27	.2083	−171.3; −120.9; −8.4	−124.8
	9 00	.2101	−171.3; −123.3; −8.4	−125.5
	9 35	.2120	−170.6; −121.8; −3.8	−123.3
	10 11	.2139	−176.1; −126.5; −1.5	−128.7
	10 52	.2161	−171.4; −124.8; −13.4	−134.2
	11 38	.2186	−171.5; −125.1	−124.2
8	3 48	.2707	−175.9; −125.5; +5.8	−128.5; −99.2
	4 28	.2728	−172.7; −129.5; −76.0; +4.2	−124.7
	5 02	.2747	−172.8; −126.4; +4.9	−117.8
	5 33	.2763	−172.8; −128.8; −57.2; +29.3	−125.5
	6 02	.2779	−176.1; −128.9; +2.5	−129.5; −98.6
	6 35	.2797	−175.3; −125.7; −69.1; +28.4	−126.4
	7 07	.2814	−178.5; −126.6; −84.9; +33.8	−128.0
	7 39	.2831	−180.1; −132.9; −85.7; +37.0	−130.4
	8h10m	.2848	−180.9; −132.1; −88.1; +4.0	−131.1
	8 41	.2864	−178.6; −125.9; +25.9	−124.3; −91.1
	9 11	.2880	−174.7; −126.7; −95.9; +32.2	−126.6
	9 44	.2898	−175.6; −133.1; −83.6; +28.1	−127.5; −95.1
	10 16	.2915	−174.8; −130.8; −78.8; +32.1	−128.2
	10 51	.2934	−175.7; −133.2; −64.0; +24.9	−129.1; −8.8
	11 22	.2951	−177.3; −133.2; −89.2; +30.4	−129.1
	11 53	.2967	−176.5; −130.1	−126.8; −92.1
9	3 56	.3485	−173.8; −126.6; −109.3; +3.9	−128.0
	4 41	.3509	−175.4; −133.7; −81.8; +21.3	−130.4; −68.6
	5 21	.3530	−175.5; −134.6; −85.8; +22.7	−132.0
	5 53	.3547	−174.7; −133.0; −81.1; +21.2	−136.6
	6 30	.3567	−174.0; −135.5; −85.1; +14.8	−132.9
	7 02	.3585	−174.0; −134.7; −78.8; +17.9	−126.7; −73.5
	7 29	.3599	−174.8; −134.7; −84.4; +28.9	−132.1
	8 00	.3616	−172.6; −131.6; −76.6; +6.8	−126.6; −63.5
	8 28	.3631	−173.3; −131.6; −78.9; +18.6	−128.2
	8 57	.3646	−175.8; −134.1; −81.4; +19.4	−126.4
	9 29	.3663	−175.8; −134.9; −79.8; +16.1	−130.0
	10 04	.3682	−175.1; −134.2; −79.1; +9.7	−128.5

TABLE 3—*Continued*

Date 1955	U.T.	Cycle and Phase (P)	Radial Velocities (Km./Sec.) from	
			He I λ3888	He I λλ3614 and 3965
July 9	10h36m	7.3699	−175.1; −134.2; −77.6; +31.8	−131.6; −59.1
	11 08	.3717	−175.2; −135.9; −80.0; +15.9	−132.7
	11 42	.3735	−173.6; −136.7; −76.9; +23.8	−133.4
10	3 48	.4254	−166.2; −129.2; −58.4; +9.8	−131.8; −84.3
	4 28	.4275	−172.6; −134.8; −59.3; +16.2	−130.3; −85.9; +28.5
	5 00	.4292	−175.7; −136.4; −59.3; +28.8	−129.1
	5 29	.4308	−175.0; −134.1; −58.6; +12.2	−130.0
	5 58	.4324	−174.2; −131.7; −57.0; +10.6	−130.3; −74.4
	6 30	.4341	−173.4; −134.1; −61.7; +30.3	−130.4; −85.2
	7 00	.4357	−175.1; −135.0; −57.9; +6.6	−130.1; −77.8; +5.8
	7 31	.4373	−173.5; −132.6; −57.9; +21.5	−126.8; −79.9
	8 05	.4392	−172.9; −131.9; −58.8; +25.4	−132.3; −87.0
	8 35	.4408	−174.4; −133.5; −57.2; +30.1	−132.7; −78.5
	9 06	.4425	−175.3; −131.3; −56.5; +31.6	−127.2; −67.8
	9 35	.4440	−173.0; −131.3; −53.4; +33.7	−130.3; −67.8
	10 10	.4459	−174.6; −132.1; −55.8; +26.0	−128.4; −74.1
	10 46	.4478	−174.6; −130.6; −61.4; +19.7	−128.4; −78.7
	11 22	.4500	−175.5; −129.9; −55.2; +47.9	−123.4; −73.3
11	11 56	.4516	−173.2; −131.5; −53.6	−132.0; −74.9
	3 56	.5031	−173.5; −135.8	−127.0
	4 50	.5060	−172.9; −132.7	−126.0; −48.0
	5 38	.5086	−174.5; −133.6; −30.6	−127.6; −35.4
	6 18	.5108	−175.3; −135.2; −47.1; +26.8	−129.1
	7 00	.5130	−169.9; −131.4; −45.6; +20.5	−138.4; −47.3
	7 38	.5151	−171.5; −132.9; −41.7; +26.0	−130.5; −49.4
	8 17	.5172	−173.2; −133.0; −36.3; +40.0	−134.0; −31.0
	8 51	.5190	−177.9; −133.8; −40.2; +21.2	−131.3
	9 29	.5210	−174.0; −134.7; −46.6; +6.1	−134.2; −34.2
	10 14	.5234	−172.6; −134.0; −53.0; +5.2	−130.7
	10 52	.5255	−174.1; −141.9; −56.9; +6.0	−133.6
	11 28	.5274	−173.5; −133.5; −57.0; +49.2	−135.0; −55.2
	12 00	.5291	−175.0; −132.5	−133.6; −29.7
Sept. 7	8 22	11.9872	−160.8; −122.3; −23.9	−125.6; −35.3
	4 43	.9915	−163.4; −124.1; −21.0	−125.0; −30.9
	6 15	.9965	−160.3; −124.1; −27.4	−122.8; −40.2
	8 00	12.0021	−165.9; −124.2; −12.5	−124.4; −7.2

departure is present when the strong line Fe II λ4233 alone is plotted.

During the eclipse the velocities from the lines of the B8 component show the effect of rotational disturbance. This effect is not symmetrical before and after mid-eclipse: it is more pronounced after mid-eclipse, a fact which could be explained in terms of an inclination of the axis of rotation of the B8 component with respect to the plane of the orbit. No evidence for changes in the velocity of the B8 star either during one night or in different cycles has been found.

THE SHELL SPECTRUM

The shell spectrum shows multiple components, which at some phases blend with the stellar lines. The radial velocities from these components at He I

λ3888 (of the triplet series) and He I λλ3614 and 3965 (of the singlet series), that is, from lines arising from metastable levels, are listed in table 3 and are plotted in figures 13 and 14. Table 4 lists the radial velocities from other He I lines of the triplet series, namely, λλ4026 and 4472, and from Hγ, which are plotted in figures 15 and 16. He I 3888 has two components with very definite and rather constant velocities, namely, about −170 and −130 km./sec. The rest of the components do not show such a constant behavior. The two components to which we have referred must arise in a shell which surrounds the entire system. The fact that the more negative component of λ3965 is observed only immediately after mid-eclipse suggests that the density of this shell is relatively higher at these phases, conceivably because of the matter which gives rise to the violet satellites, part of which probably escapes from the system through the external Lagrangian point in front of the secondary component of the system. The different behavior of the more negative component of the He I lines λ3888 and λ3965 also suggests that the shell expands outward.

The shell velocities of λ3888 Mount Wilson coudé plates taken by Struve in 1952 and 1953 are not appreciably different from those obtained in 1955. Neither are there any very large differences in the intensities of the lines. It would seem that the system has been losing mass at a rather constant rate in the past few years.

THE SATELLITE LINES

The radial velocities from the satellite lines are listed in tables 5 and 6 and are plotted in figure 17. In the plot of the violet satellite lines only the velocities from the narrow absorptions have been entered. The velocities from the red satellites tend to be algebraically larger as we approach mid-eclipse, and the opposite seems to be true for the violet satellites.

THE EMISSION LINES

Shifts in the position of the emissions are roughly opposite in phase with respect to the absorption lines of the B8 star. The relatively narrow, strong "peaks" of the emission structures have been measured for radial velocity and the results for He I 3888 and 4472 and for Hγ are listed in table 7 and are plotted in figures 18, 19 and 20. These velocities are in phase with those expected for the unobserved lines of the secondary component of the system; but the γ-velocity is about +130 km./sec. larger than that of the B8 component. The semi-amplitude is of the order of 30 km./sec.

INTERSTELLAR CA II

The mean velocity of the interstellar H and K lines from 191 plates is − 15.2 ± 0.2 km./sec.

TABLE 4

RADIAL VELOCITIES FROM SHELL LINES OF HE I AND H

Date 1955	U.T.	Cycle and Phase (P)	He I λλ4472 and 4026	Hγ
Apr. 6	10^h38^m	0.0999	−187.2	−192.0
	12 16	.1051	−185.7; −110.4	−184.0
7	9 58	.1748	−116.0	−79.7
	10 55	.1781	−169.2; −112.8	−112.6
	11 56	.1814	−110.8	−18.3
8	9 51	.2520	−84.3	
	10 38	.2546	−104.3; −44.8; −6.0	−91.7; −18.4
	11 24	.2570	−86.7	−92.4; −40.3; +38.6
	12 11	.2595	−94.1; −33.2	−97.5; −41.1; +32.2
May 5	10 00	2.3408	−35.0; +48.6	−166.7; −41.3; +58.1
	10 49	.3434	−52.7; +15.7; +67.0	−38.6; +23.5
	11 42	.3462	−55.4; +21.1	−37.2; +37.6
6	7 03	.4086	−28.7; +35.6	−25.8
	7 59	.4116	−28.0; +42.5	−28.0; +45.4
	9 00	.4149	−23.2; +51.0	−25.9; +39.7
9	7 25	.6418	−69.2	−72.0
	8 27	.6451	−72.1; +4.1	−74.1
	9 28	.6484	−74.4	−75.6
	10 30	.6517	−73.7; +8.4	−82.1
	11 36	.6553	−73.9	−84.3
June 3	5 21	4.5687	−141.6; −91.0	−142.7; −70.9; +15.1
	6 32	.5725	−142.4; −91.8	−123.8; +5.2
	7 26	.5754	−141.2; −86.6	−136.6
	8 40	.5794	−135.0; −66.4	−124.0
	9 43	.5828	−138.9; −79.7	−172.0; −119.9
	11 47	.5894	−131.3; −75.2	−157.4; −108.8
4	6 30	.6498	−96.8	−118.4
	7 34	.6532	−95.1; −3.1	−116.4
	8 34	.6564	−106.7	−110.1; −53.7
	9 32	.6595	−142.4; −81.6	−110.2; −34.1
	10 25	.6624	−146.9; −103.9	−160.4; −111.0; −64.6
	11 18	.6652	−98.0	−166.1; −118.2; −35.7
5	7 02	.7288	−97.2; −23.8	−139.8; −51.0
	8 06	.7323	−149.2; −73.5	−132.9; −42.0
	9 06	.7355	−136.3; −73.4; −0.2	−137.2; −47.0
	10 05	.7386	−77.6	−130.3; −43.6
	11 11	.7422	−120.4; −37.0	
6	7 10	.8066	−104.0; −48.7; −5.8	
	8 12	.8099	−114.6; −64.0	−111.2; −59.8; +3.0
	9 10	.8130	−117.8; −63.2; −9.8	−119.1; −59.2; −11.2
	10 10	.8163	−107.6; −40.1	
	11 12	.8196	−108.0; −2.4	−99.5; −54.4
	11 58	.8210	−114.2	−117.3; −48.2
7	6 00	.8802	−121.4; −44.8; +18.1	−115.5; −2.8
	6 46	.8826	−120.8; −55.1; +13.2	−121.9; +14.1
	7 34	.8852	−115.0	−121.3; +4.2
	8 29	.8882	−119.5	−117.9; −36.4; +29.4
	9 15	.8906	−122.3; +5.5	−118.7; +4.7
	9 46	.8923	−126.0; +32.8	−122.9; +1.1
	10 18	.8940	−122.2	−126.5; −19.4
	10 51	.8958	−139.0; +12.2	−125.1; −6.7
	11 21	.8974	−121.8	−121.7; −14.6
	11 52	.8991	−122.0	−123.1
8	5 17	.9552	−129.3; −71.2; −8.2	
	6 21	.9586	−135.4; −67.9	−75.7; +0.4
	7 12	.9614	−138.2; −42.4; +20.5	−111.6; −36.9; +23.7
	8 04	.9642	−146.8; −75.0; +4.8	−112.4; −33.5; +15.8
	9 05	.9675	−134.0; −69.8; +13.5	−116.0; −47.0; +9.4
	10 15	.9712	−119.1; −70.5; +1.8	
	11 31	.9753	−137.0; −75.0; −9.8	−81.7; +16.3
9	4 56	5.0314	−138.5; −92.4	−110.3
	6 31	.0365	−121.9	−82.9
	7 31	.0397	−107.7	−99.2
	8 34	.0411	−128.3	−107.7
	10 32	.0415	−139.4	−105.9
	11 16	.0518	−78.5	−110.8
10	5 04	.1092	−170.6	−154.3
	5 45	.1114	−162.4	−158.5
	6 21	.1133	−162.4	−168.6
	6 47	.1147	−166.3	−158.7
	7 09	.1159	−170.0; −104.0	−178.6
	7 30	.1170	−167.0	−146.6
	7 51	.1182	−170.8	−164.4
	8 12	5.1193	−161.8	−157.3
	8 33	.1204	−164.9	−156.6
	8 54	.1215	−169.0	−163.1
	9 25	.1232	−163.4	−169.5
	9 56	.1249	−146.5	−165.4
	10 27	.1265	−155.2	−159.0
	10 59	.1282	−154.0	−159.7
	11 30	.1299	−155.2	−154.1
	11 58	.1314	−155.7	−150.5
July 4	4 03	6.9621	−100.4; −25.2	−79.6; −15.4
	5 00	.9652	−105.8; −19.9	−171.9; −15.4
	6 16	.9693	−119.1; −70.7; −18.2	−122.6; −79.0; −23.3
	7 34	.9735	−121.5; −28.5; +18.6	−129.2; −27.7
	8 25	.9762	−108.8; −27.2	−125.7; −31.9
	9 13	.9788	−115.5; −24.8	−122.2; −70.8; −21.5

Date 1955	U.T.	Cycle and Phase (P)	He I λλ4472 and 4026	Hγ
July 4	10^h02^m	6.9814	−110.8; −22.1	−97.7; −13.8
	10 47	.9838	−114.2; −57.7; −10.1	−127.3; −70.2; −23.7
	11 44	.9869	−122.2; −71.3; −15.9	−105.5; −29.4
5	3 55	7.0390	−120.1	−124.2; −59.3; −4.3
	4 49	.0419	−124.0	−122.8
	5 50	.0452	−125.4; −60.0	−119.4
	6 41	.0480	−121.1	−118.0
	7 36	.0509	−123.4; −81.1	−110.4
	8 37	.0542	−116.6	−118.2
	9 28	.0569	−113.3	−112.0
	10 27	.0601	−117.1	−111.4
6	3 52	.1162	−70.8	−52.6
	4 40	.1188	−57.5	−54.8
	5 21	.1210	−68.3	−50.6
	6 02	.1232	−63.4	−54.9
	6 43	.1254	−66.8; +34.2	−52.1; +25.5
	7 27	.1278	−171.0; −60.4; +39.4	−167.0; −57.8; +24.0
	8 08	.1300	−60.5; +34.2	−65.7
	8 49	.1322	−58.2; +31.2	−57.9
	9 30	.1344	−161.6; −64.8; +37.7	−46.0
	10 16	.1368	−171.3; −61.4; +26.2	−163.1; −55.3
	11 15	.1400	−73.0	−47.5
	12 02	.1425	−72.3	
7	3 53	.1936	−125.0; −27.6	−120.5; −36.0
	4 34	.1958	−130.4; −32.6	−122.6; −34.5; +52.9
	5 15	.1980	−124.2; −35.6	−115.7; −36.1
	5 56	.2002	−126.8; −23.0	−115.8; −28.4
	6 35	.2023	−123.4; −39.2	−120.7; −15.7
	7 16	.2045	−121.8; −8.0	−121.5; −25.0
	7 52	.2064	−122.7; −42.7	−122.9; −43.3
	8 27	.2083	−120.0; +4.0	−118.1; −43.4
	9 00	.2101	−123.6; +4.7	−119.5; −7.5
	9 35	.2120	−119.2; −4.3	−114.7; −25.2
	10 11	.2139	−118.6; −20.0	−122.4; −24.5
	10 52	.2161	−124.0; −17.7	−117.6; −9.8
	11 38	.2186	−122.5; −1.9	−121.1; −16.1
8	3 48	.2707	−107.8; −1.4	−105.3; +4.6
	4 28	.2728	−111.6; +17.3	−101.8; +12.4
	5 02	.2747	−111.9; +17.1	−103.3; +10.2
	5 33	.2763	−111.8; +10.3	−103.3; +21.4
	6 02	.2779	−107.4; +23.5	−99.2; +23.5
	6 35	.2797	−107.8; −0.6	−102.0; +7.9
	7 07	.2814	−108.0; +10.8	−98.6
	7 39	.2831	−106.2; +18.0	−96.5; +0.8
	8 10	.2848	−111.8; −6.8	−97.9; +33.2
	8 41	.2864	−107.0; +12.4	−110.0; +20.4
	9 11	.2880	−102.1; −2.3	−94.4; −2.8
	9 44	.2898	−100.2; +1.5	−100.9; +3.4
	10 16	.2915	−102.8; +3.3	−103.7; +3.4
	10 51	.2934	−105.0; +14.2	−98.2; +4.0
	11 22	.2951	−104.9; +18.4	−100.3; +6.1
	11 53	.2967	−101.3; +21.0	−96.8; +16.0
9	3 56	.3485	−77.5; +7.4	−79.5; +14.2
	4 41	.3509	−67.6; +9.7	−61.2; +27.6
	5 21	.3530	−82.7; +26.7	−70.5; +30.3
	5 53	.3547	−74.5; +16.2	−72.6; +14.1
	6 30	.3567	−69.4; +10.8	−72.7; +10.5
	7 02	.3585	−76.7; +31.4	−74.1; +9.8
	7 29	.3599	−72.6; +22.8	−55.8; +28.8
	3 00	.3616	−133.6; −76.1; +23.8	−70.7; +13.2
	8 28	.3631	−139.7; −65.2; +21.4	−144.7; −75.6; +4.0
	8 57	.3646	−141.9; −65.5; +12.0	−58.8; +20.1
	8 29	.3663	−70.7; +30.4	−141.3; −56.0; +30.0
	10 04	.3682	−132.4; −75.6; +20.8	−53.3; +17.2
	10 36	.3699	−74.9; +21.5	
	11 08	.3717	−75.7; +26.9	−114.0; −55.5; +31.2
	11 42	.3735	−72.3; −10.4	−55.2; +24.9
	3 48	.4254	−57.5; +28.6	−59.3; +4.1
	4 28	.4275	−85.9; +28.5	−44.6; +27.3
	5 00	.4292	−70.6; +13.5	−70.6; +13.5
	5 29	.4308	−50.9	−47.5
	5 58	.4324	−85.1; +16.1	−47.5
	6 30	.4341	−65.3; +20.9	−92.6; −13.7
	7 00	.4357	−77.7; +5.8	−53.3
	7 31	.4373	−46.2; +6.5	−39.2; +12.3
	8 05	.4392	−75.0; +5.0	−71.5; +11.5
	8 35	.4408	−73.0	−59.0; −0.5
	9 06	.4425	−82.7; −47.1; +4.9	−76.0; −4.8
	9 35	.4440	−66.9; +30.2	−4.1
	10 10	.4459	−68.4; −9.6	−57.1; +4.9
	10 46	.4478	−115.6; −67.0	−122.6; −62.7
	11 22	.4500		−34.6
11	11 28	.5274	−141.3	
Sept. 7	3 22	11.9872	−117.7	−90.5
	4 43	.9915	−118.4; −17.3	−99.9; −66.1
	6 15	.9965	−117.2; −50.4	−85.2
	8 00	12.0021	−117.4; −67.2	−101.5; −47.8

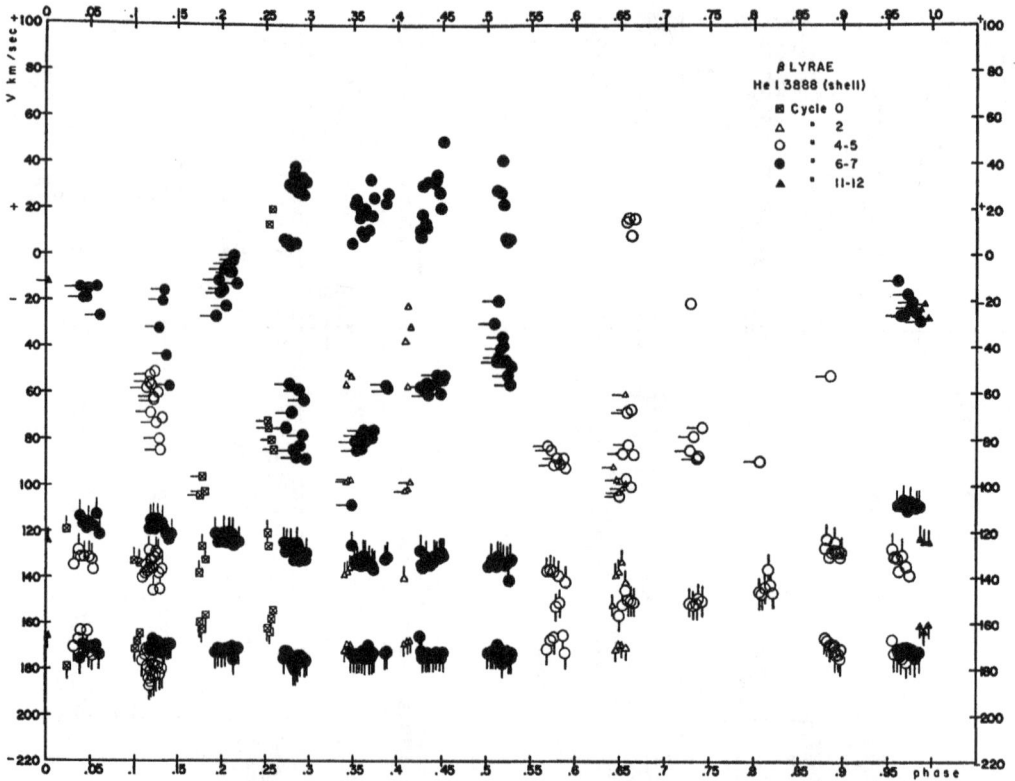

Fig. 13. Radial velocities from the shell (triplet) line at He I λ3888. The "tails" distinguish the groupings which the different components suggest. Two components yield rather constant velocities of about −170 and −130 km./sec. The rest of the components do not behave so regularly and show large differences in different cycles.

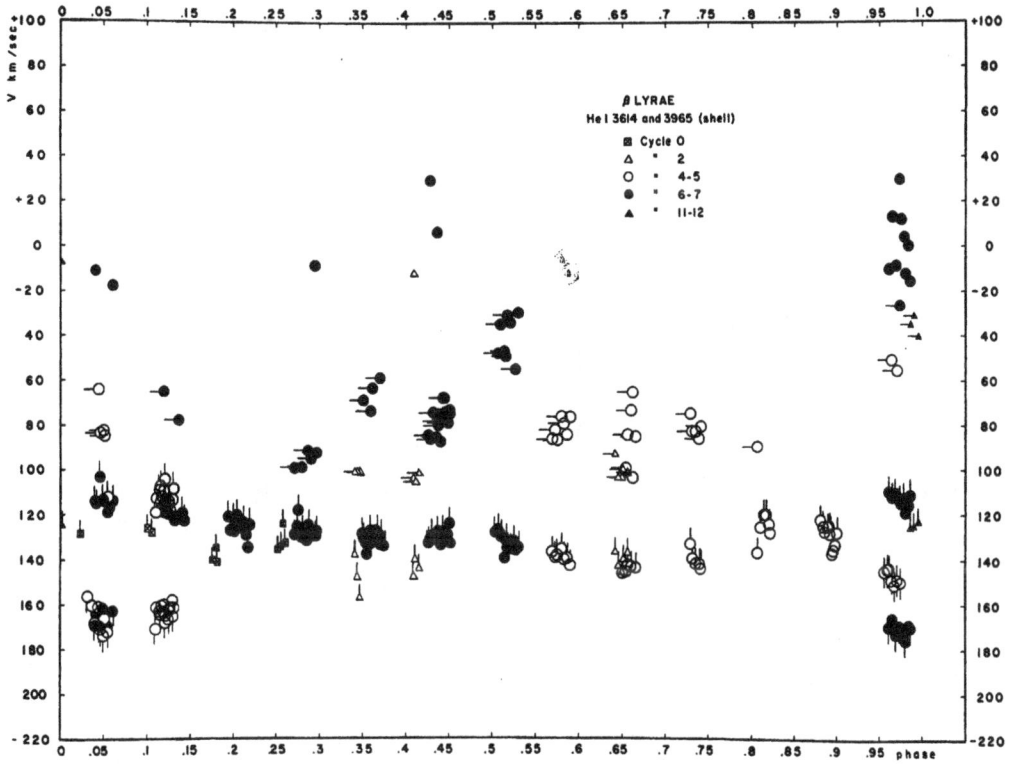

FIG. 14. Radial velocities from the shell (singlet) lines at He I λλ3614 and 3965. The "tails" distinguish the groupings which the different components suggest.

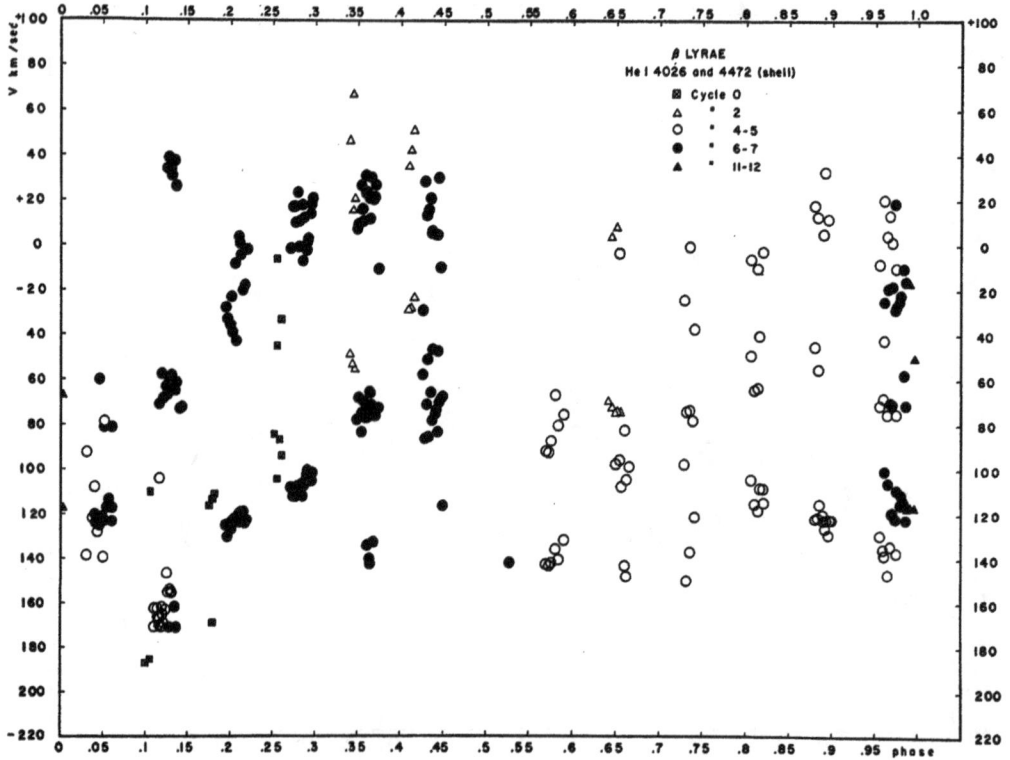

FIG. 15.   Radial velocities from the shell (triplet) lines at He I λλ4026 and 4472.   Unlike the cases in figures 13 and 14, it is difficult to attribute the velocities from the different components at different phases, to definite groupings.

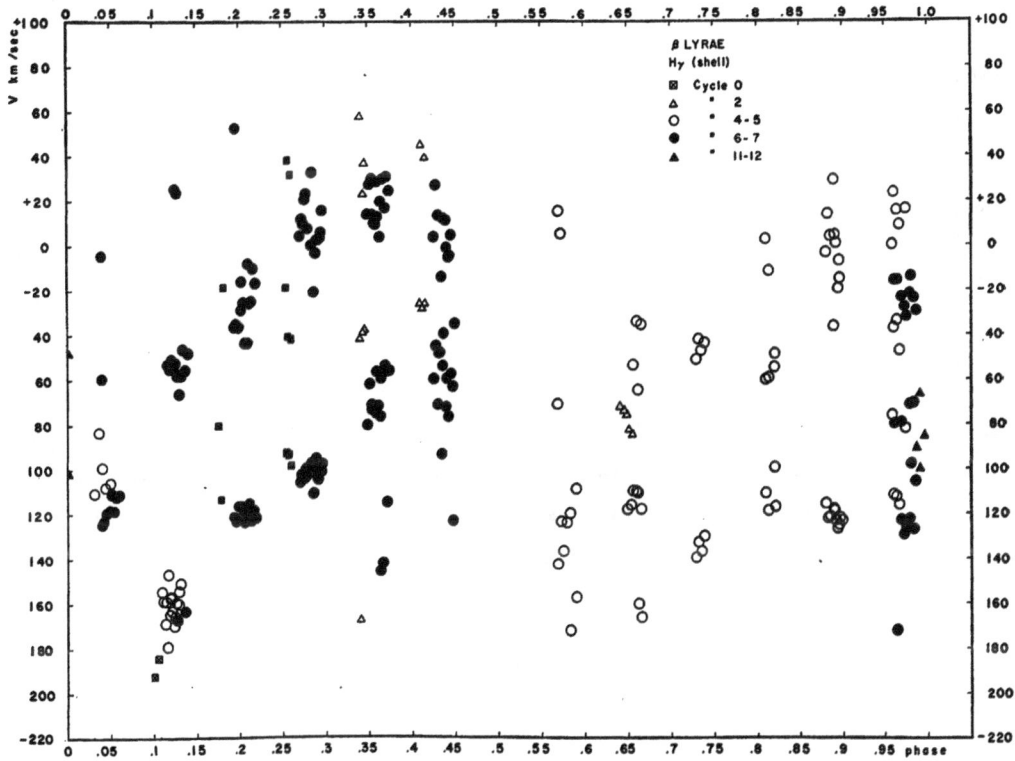

FIG. 16.   Radial velocities from the shell line at Hγ.   Unlike the cases in Figures 13 and 14, it is difficult to attribute the velocities from the different components at different phases, to definite groupings.

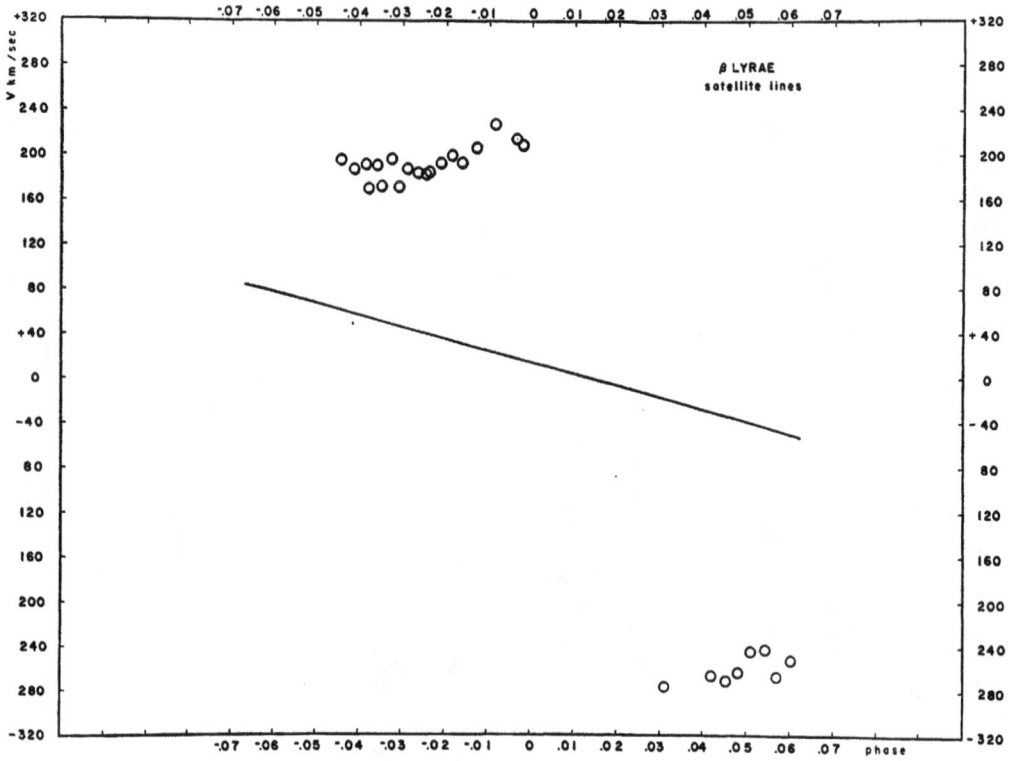

FIG. 17.　Radial velocities from the satellite lines and velocity curve from Si II at the relevant phase interval.

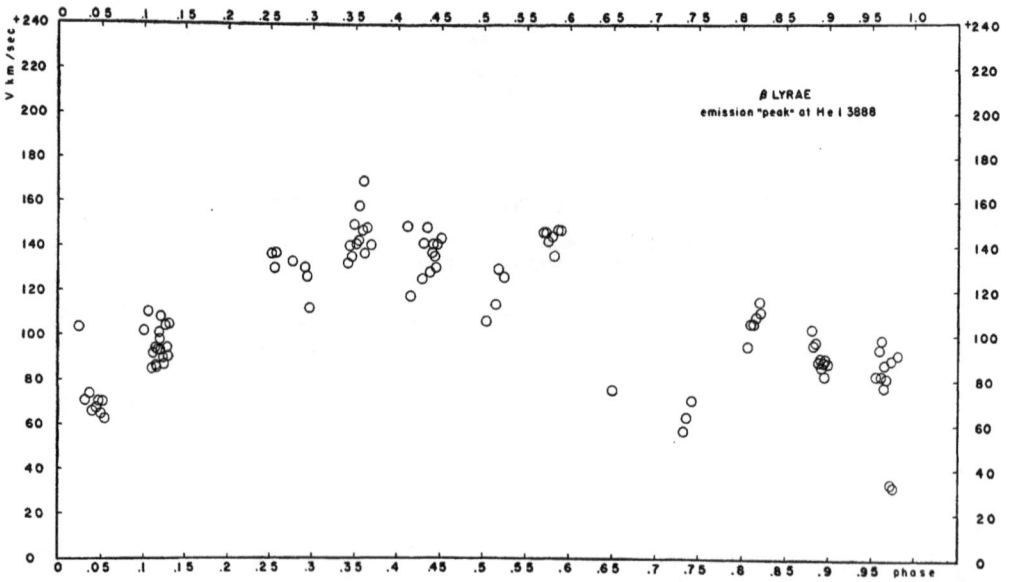

FIG. 18.　Radial velocities from the emission "peak" at He I λ3888.

TABLE 5
RADIAL VELOCITIES FROM THE RED SATELLITE LINES

Date 1955	U.T.	Cycle and Phase (P)	Velocity (Km./Sec.)
June 8	5h17m	4.9552	+176.6
	6 21	.9586	189.2
	7 12	.9614	192.5
	8 04	.9642	191.8
	9 05	.9675	197.5
	10 15	.9712	189.7
	11 31	.9753	184.1
July 4	4 03	6.9621	171.4
	5 00	.9652	173.5
	6 16	.9693	172.9
	7 34	.9735	185.6
	8 25	.9762	186.0
	9 13	.9788	193.6
	10 02	.9814	200.8
	10 47	.9838	194.5
Sept. 7	3 22	11.9872	207.8
	4 43	.9915	228.8
	6 15	.9965	215.8
	8 00	12.0021	+210.3

THE LEAST–SQUARES SOLUTION

Following the conventional notation [16], we have for the radial velocity V of the component of a spectroscopic binary in a heliocentric system of reference:

$$V = \gamma + K_1[e \cos \omega + \cos (v + \omega)]. \quad (1)$$

The equations of condition for the least-squares solution are derived by taking the variations of equation (1) as was first done by Lehmann-Filhés [17]. Schlesinger [18] rearranged the equations of condition in a form most suitable for practical application and also provided the necessary tables [19] to facilitate the actual computation of the orbital elements. The standard form developed by Schlesinger is still

TABLE 6
RADIAL VELOCITIES FROM THE VIOLET SATELLITE LINES

Date 1955	U.T.	Cycle and Phase (P)	Velocity (Km./Sec.)		
			Narrow Absorption	Edges of Broad Absorption ($H\delta$)	
June 9	6h31m	5.0365		−373	−163
	7 31	.0397		−385	−155
	8 34	.0431		−408	−144
	9 40	.0467		−398	−152
	10 32	.0495		−406	−138
	11 16	.0518		−360	−115
July 5	3 55	7.0390	−276.5		
	4 49	.0419	−266.6	−356	−153
	5 50	.0452	−271.4	−435	−142
	6 41	.0480	−263.6		
	7 36	.0509	−244.9	−424	−194
	8 37	.0542	−242.7	−360	−159
	9 28	.0569	−266.4		
	10 27	.0601	−252.4	−339	−186

often used. However, some difficulty arises in Schlesinger's method when e, the eccentricity of the orbit, is very small. This difficulty is obvious, because the time of periastron passage, T, and the longitude of the periastron, ω, are not well-determined quantities for orbits of small e and they lose their meaning completely when $e = 0$. Many investigators have arbitrarily fixed T or ω in applying Schlesinger's method to orbits of very small eccentricity and have thereby obtained misleadingly small values for the mean errors of the orbital elements. Luyten [20] proposed in 1936 the replacement of T by the time of nodal passage, which remains significant even in a circular orbit, and Sterne [21] formulated in 1941 the equations of condition in the least-squares solution for the case of small eccentricity. Therefore, two methods are now available for computing the orbital elements: one for large e and the other for small e. These are the methods which are best suited to a computer to whom only a desk machine is available.

The introduction of the electronic digital computers of high speed changes the picture. It renders Schlesinger's procedure with the use of his many tables unnecessary, because the machine can compute in a small fraction of a second all the quantities that are involved in the solution. The subdivision into two cases in the least-squares solution according to the value of e can also be avoided. Consequently, we shall reformulate the equations of condition in such a way that they will provide most profitably the basis for writing (on any electronic digital computer) a subroutine which will determine automatically the orbital elements of a spectroscopic binary after the radial velocities have been fed into the machine. In essence, the present formulation is a generalization of Sterne's method for orbits of very small eccentricity to a form which holds true for orbits of any eccentricity.

In the case of Beta Lyrae and also of most other spectroscopic binaries, the period, P, is well determined by several series of observations which are separated by time intervals which are long compared to the period. Therefore, in the least-squares solution, we shall omit the variation in P, leaving altogether five independent orbital elements that can be freely varied in the solution. From equation (1)

$$\delta V = \delta\gamma + [\cos (v + \omega) + e \cos \omega]\delta K_1$$
$$+ K_1\left(\cos \omega - \frac{\sin (v + \omega) \sin v(2 + e \cos v)}{1 - e^2}\right)\delta e$$
$$- K_1[\sin (v + \omega) + e \sin \omega]\delta\omega$$
$$+ \frac{K_1\mu}{(1 - e^2)^{\frac{3}{2}}} \sin (v + \omega)(1 + e \cos v)^2\delta T. \quad (2)$$

Following Sterne [21], we introduce the "epoch of the mean longitude," T_0, at which the mean longitude

TABLE 7

RADIAL VELOCITIES FROM THE EMISSION "PEAKS"

Date 1955	U.T.	Cycle and Phase (P)	HeI λ3888	HeI λ4472	Hγ	Date 1955	U.T.	Cycle and Phase (P)	HeI λ3888	HeI λ4472	Hγ
April						June					
5	10h57m	0.0235	+103.3			9	11h16m	5.0518	+ 70.4	+ 78.8	—
6	10 38	.0999	+101.8	+133.8	+162.5		11 52	.0538	+ 62.5	+ 99.9	—
	12 16	.1051	+110.4	+141.3	+155.4	10	5 04	.1092	+ 84.8	+ 98.8	—
7	10 55	.1781	—	+160.4	—		5 45	.1114	+ 91.9	+128.2	+143.8
8	9 51	.2520	+136.5	+152.3	—		6 21	.1133	+ 94.1	—	+141.3
	10 38	.2546	+130.1	+162.5	+154.9		6 47	.1147	+ 86.3	+139.7	+149.2
	11 24	.2570	+137.1	+161.8	+188.8		7 09	.1159	+ 85.5	+138.4	+137.1
	12 11	.2595	—	—	+182.3		7 30	.1170	+ 93.4	+131.5	+132.8
May							7 51	.1182	+101.0	+143.0	+146.2
5	10 00	2.3408	+132.3	+181.3	+161.7		8 12	.1193	+ 98.0	+131.4	+143.4
	10 49	.3434	+140.1	—	+146.8		8 33	.1204	+ 93.3	+147.8	+134.8
	11 42	.3462	+135.3	+161.4	—		8 54	.1215	+108.1	+136.1	+155.4
6	7 03	.4086	—	+149.8	—		9 25	.1232	+ 90.0	+118.3	+151.1
	7 59	.4116	+148.8	—	+157.4		9 56	.1249	+ 86.8	+119.6	+143.9
	9 00	.4149	+117.8	+135.4	+128.5		10 27	.1265	+104.1	+138.1	+148.9
June							10 59	.1282	+ 94.5	+135.2	—
3	5 21	4.5687	+146.3	+141.5	—		11 30	.1299	+ 90.6	+139.3	+138.8
	6 32	.5725	+146.2	+107.9	+119.3		11 58	.1314	+104.8	+136.6	+143.8
	7 26	.5754	+142.1	+106.4	+114.3	July					
	8 40	.5794	+144.4	+109.8	—	4	4 03	6.9621	+ 98.2	—	—
	9 43	.5828	+135.7	—	+128.6		5 00	.9652	+ 87.2	—	—
	10 46	.5862	+147.4	+137.5	—		7 34	.9735	+ 89.2	—	—
	11 47	.5894	+147.3	+133.4	+135.8		10 02	.9814	+ 91.4	—	—
4	6 30	.6498	+ 76.0	—	—	7	9 00	7.2101	—	+131.0	—
5	8 06	.7323	+ 57.5	+ 81.3	—		9 35	.2120	—	+146.7	—
	9 06	.7355	+ 63.7	+ 85.3	—	.8	5 33	.2763	+133.1	—	—
	10 05	.7386	—	+ 83.2	—		6 02	.2779	—	+161.0	+167.9
	11 11	.7422	+ 71.3	—	—		9 44	.2898	—	+132.6	—
6	7 10	.8066	+ 95.1	+ 93.3	+102.2		10 16	.2915	+130.4	+131.2	—
	8 12	.8099	+105.3	+ 98.9	+110.8		10 51	.2934	+126.4	+145.5	—
	9 10	.8130	+105.2	+100.9	+ 93.8		11 22	.2951	—	—	+157.0
	10 10	.8163	+108.2	+106.9	+ 92.3		11 53	.2967	+112.2	—	—
	11 12	.8196	+115.2	+ 98.5	+102.7	9	3 56	.3485	+149.5	—	—
	11 58	.8210	+110.4	+ 99.2	+ 92.8		4 41	.3509	+140.8	—	+148.8
7	6 00	.8802	+102.8	+ 92.0	+ 93.8		5 21	.3530	+142.3	—	—
	6 46	.8826	+ 95.7	+ 87.2	—		5 53	.3547	+158.0	—	+145.9
	7 34	.8852	+ 97.2	+ 87.1	—		7 02	.3585	+146.9	—	+145.1
	8 29	.8882	+ 88.4	+ 78.8	+ 89.1		7 29	.3599	+168.9	+152.4	+150.7
	9 15	.8906	+ 89.9	—	—		8 00	.3616	+136.6	+156.5	—
	9 46	.8923	+ 86.0	—	—		8 28	.3631	—	—	+138.6
	10 18	.8940	+ 88.2	+ 76.5	—		8 57	.3646	+148.3	—	—
	10 51	.8958	+ 82.0	—	—		10 04	.3682	+140.5	—	—
	11 21	.8974	+ 89.7	—	—	10	3 48	.4254	—	+133.3	—
	11 52	.8991	+ 87.4	+ 89.0	—		4 28	.4275	+125.6	—	—
8	5 17	.9552	+ 82.2	+ 82.2	+ 84.4		5 00	.4292	+141.3	—	—
	6 21	.9586	+ 93.9	+ 86.2	+ 85.0		6 30	.4341	+148.3	—	—
	7 12	.9614	+ 82.1	+ 83.5	+ 86.4		7 31	.4373	+128.5	—	—
	8 04	.9642	+ 77.2	+ 77.2	+ 81.4		8 05	.4392	+137.1	—	—
	9 05	.9675	+ 81.1	+ 89.4	+ 86.2		8 35	.4408	+141.0	—	—
	10 15	.9712	+ 33.8	+ 81.8	+ 81.2		9 06	.4425	+135.4	—	—
	11 31	.9753	+ 32.1	+ 78.3	+ 82.5		9 35	.4440	+130.5	—	—
9	4 56	5.0314	+ 70.9	+ 73.1	—		10 10	.4459	+140.8	—	—
	6 31	.0365	+ 74.0	+ 77.1	—		11 22	.4500	+143.8	—	—
	7 31	.0397	+ 66.0	+ 75.6	—	11	3 56	.5031	+106.5	+100.8	+103.2
	8 34	.0431	+ 67.5	—	—		7 00	.5130	+114.1	—	—
	9 40	.0467	+ 70.5	+ 78.2	—		8 17	.5172	+129.9	—	—
	10 32	.0495	+ 65.1	+ 72.6	—		10 14	.5234	+126.3	—	—

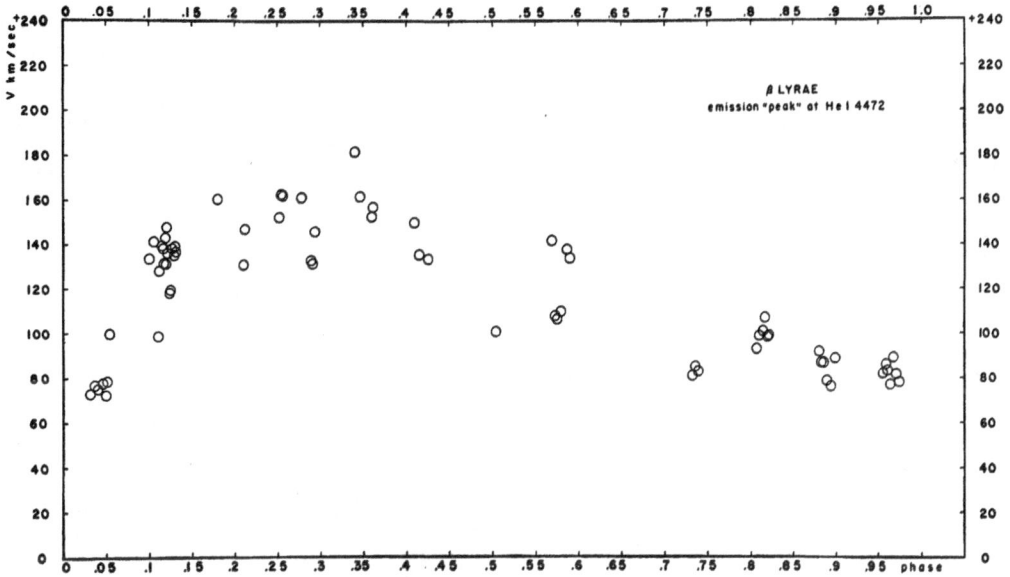

FIG. 19.   Radial velocities from the emission "peak" at He I λ4472.

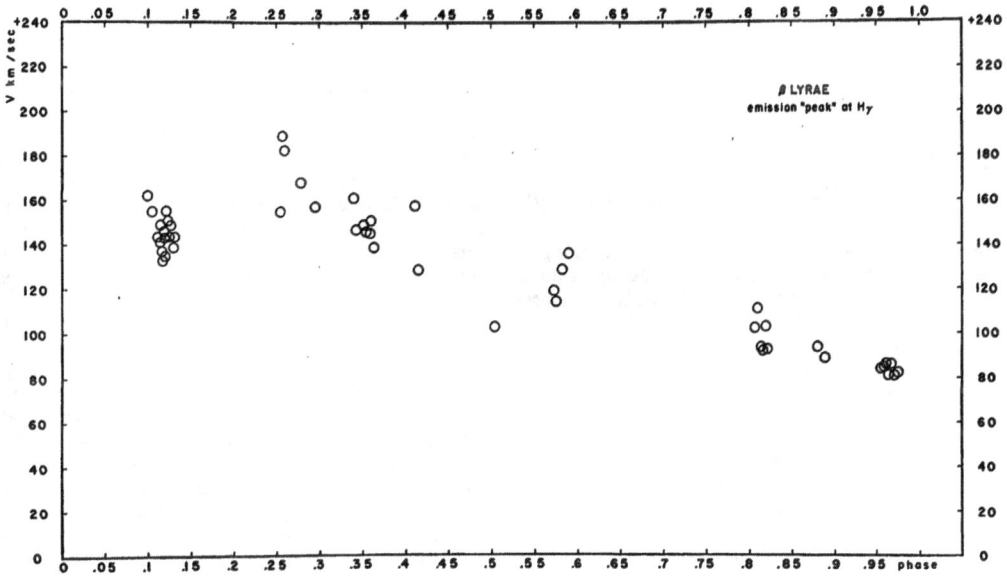

FIG. 20.   Radial velocities from the emission "peak" at Hγ.

$\omega + M$ is zero. Obviously, T_0 remains determinate in the limit of vanishing e. Since the mean anomaly M is given by

$$M = \mu(t - T), \tag{3}$$

where

$$\mu = \frac{2\pi}{P} \tag{4}$$

we have

$$T = T_0 + \frac{\omega}{\mu}. \tag{5}$$

Also we introduce two new variables ξ and η such that

$$\xi = e \cos \omega \quad \text{and} \quad \eta = e \sin \omega. \tag{6}$$

Equations (5) and (6) transform the set of three variables, T, e and ω into an equivalent set of three new variables: T_0, ξ and η; the latter will be used in the least-squares solution, because they are well-defined quantities for all orbits.

By differentiating equations (5) and (6), solving for δT, δe, and $\delta \omega$ in terms of δT_0, $\delta \xi$, and $\delta \eta$, and substituting them into equation (2), we obtain the final equations of condition:

$$A_1 \delta\gamma + A_2 \delta K_1 + A_3 \delta\xi + A_4 \delta\eta + A_5 \delta T_0 = A_6, \tag{7}$$

where

$$
\begin{aligned}
A_1 &= (1 - e^2)^{\frac{1}{2}} \\
A_2 &= (1 - e^2)^{\frac{1}{2}} [\cos(v + \omega) + e \cos \omega] \\
A_3 &= K_1 [(1 - e^2)^{\frac{1}{2}} - \alpha \cos \omega - \beta \sin \omega] \\
A_4 &= K_1 [\beta \cos \omega - \alpha \sin \omega] \\
A_5 &= K_1 \mu \sin(v + \omega)(1 + e \cos v)^2 \\
A_6 &= (1 - e^2)^{\frac{1}{2}} \delta V
\end{aligned}
\tag{8}
$$

and

$$\alpha = (1 - e^2)^{\frac{1}{2}} \sin(v + \omega) \sin v (2 + e \cos v)$$

$$\beta = \frac{(1 + e \cos v)^2 - (1 - e^2)^{\frac{1}{2}}}{e} \sin(v + \omega). \tag{9}$$

Equation (7) has been used in the programming of the solution on the machine. Once the equation of condition is derived, we can perform the least-squares solution according to the standard procedure.

We have set up the program, on the IBM electronic data processing machine 701, of the least-squares solution of a spectroscopic binary orbit as a subroutine. The only input required in this routine, besides the observed data (i.e., time, t_k, and radial velocity, v_k), is (1) the order of magnitude of γ and of K_1, and (2) the approximate time at which the radial velocity curve reaches its maximum. In this way we can omit the determination of a preliminary orbit, since both the order of magnitude of γ and K_1 and the time of nodal passage can be estimated from the observed velocity curve.

For each observation, say, the nth, the machine solves the Kepler equation internally by Newton's method of successive approximations, computes the products $A_i{}^{(n)} A_j{}^{(n)}$ (i, j from 1 to 6), adds them to the previous sums $\sum_{k=1}^{n-1} A_i{}^{(k)} A_j{}^{(k)}$ for form $\sum_{k=1}^{n} A_i{}^{(k)} A_j{}^{(k)}$ and then proceeds to the next observation. It takes either a part of, or the entire, input data (t_k, v_k) according to the command fed into the machine at the start of the operation and computes the orbital elements. In other words, we can feed into the machine at the beginning a total number of N' observations and use only N ($\leq N'$) observations for computing the orbit, while at the end the computer will print the $O - C$ for all N' observations based upon the final solution. After each least-squares solution, the change in the orbital elements $\delta\gamma$, δK_1, $\delta\xi$, $\delta\eta$ and δT_0 are compared internally with the mean errors of the corresponding quantities: $\Delta\gamma$, ΔK_1, $\Delta\xi$, $\Delta\eta$ and ΔT_0. If one more of the following inequalities

$$\delta\gamma < \Delta\gamma, \quad \delta K_1 < \Delta K_1, \quad \delta\xi < \Delta\xi,$$

$$\delta\eta < \Delta\eta, \quad \delta T_0 < \Delta T_0 \tag{10}$$

are not satisfied, another least-squares solution is performed using the orbital elements determined by the first solution as the preliminary elements for the second solution. This process repeats itself several times until all inequalities in (10) are satisfied. For each individual least-squares solution, the computer prints: (1) the coefficients of the normal equations (i.e., $\sum_{k=1}^{N} A_i{}^{(k)} A_j{}^{(k)}$), (2) $\delta\gamma$, δK_1, $\delta\xi$, $\delta\eta$, δT_0; $\Delta\gamma$, ΔK_1, $\Delta\xi$, $\Delta\eta$, ΔT_0; the auxiliary orbital elements γ, K_1, ξ, η, T_0; the mean error per plate; and (3) the orbital elements γ, K_1, e, ω, T, T_0 and their corresponding mean errors. After the solution which satisfies the inequalities given by (10) is derived, the computer automatically shifts to the computation of the $O - C$ of all observed points and prints them. It also calculates and prints an ephemeris based upon the final orbital elements at thirty-three equal intervals of the eccentric anomaly. Finally, $a_1 \sin i$ in kilometers and the mass function in units of the solar mass are calculated and printed. All these calculations and printings proceed continuously; hence it takes only a few minutes to perform these calculations even when several hundred observations are used to derive the elements. Usually, the machine iterates the least-squares solution about three to six times before all inequalities in (10) are satisfied. The fixed binary-point system has been used throughout the program.

Using the IBM 701 computer, we have computed, with the subroutine just described, several sets of the orbital elements derived from the observational data taken at different epochs since the turn of the century. The results are given in table 8, together with the sources of the observational data. Because of the well-known effect of rotational distortion [22] during primary eclipse, all observational data taken between $0.85\ P$ and $0.15\ P$ were excluded in computing the orbital elements. When these new sets of orbital elements are compared with those derived previously

TABLE 8

ORBITAL ELEMENTS OF β LYRAE FROM RADIAL VELOCITIES DERIVED BY SEVERAL INVESTIGATORS

	Belopolsky [1]	Baxandall [2]	Rossiter [3]	Sherman [4]
γ	-14.6 ± 5.9 km./sec.	-16.8 ± 3.4 km./sec.	-18.9 ± 0.4 km./sec.	-20.0 ± 1.1 km./sec.
K	180.7 ± 4.8 km./sec.	188.5 ± 2.0 km./sec.	183.9 ± 0.3 km./sec.	183.6 ± 0.7 km./sec.
e	0.026 ± 0.032	0.020 ± 0.013	0.013 ± 0.003	0.018 ± 0.005
ω	$311° \pm 74°$	$107° \pm 53°$	$327° \pm 10°$	$296° \pm 22°$
$T_0{}^*$	10.285 ± 0.110 days	9.611 ± 0.060 days	9.797 ± 0.009 days	9.655 ± 0.021 days
T^*	8.538 ± 2.642 days	0.530 ± 1.908 days	8.601 ± 0.350 days	7.363 ± 0.787 days
$a \sin i$	—	—	32.7×10^6 km.	—
$f(\mathfrak{M})$	—	—	$8.4 \odot$	—
m.e. per plate	± 11.9 km./sec.	± 10.1 km./sec.	± 3.9 km./sec.	± 4.4 km./sec.
No. of observations	20	55	301	77

* After primary minimum.

References:

1. Belopolsky, A.: *Astrophys. Jour.* **6**: 328–337, 1897.
2. Baxandall, F. E.: *Ann. Solar Phys. Obs., Cambridge* **2** (part 1): 1–24, 1930.
3. Rossiter, R. A.: *Pub. Obs. Univ. Michigan* **5**: 67–90, 1934.
4. Sherman, F.: *Astrophys. Jour.* **94**: 368–374, 1941.

from the same data by the early investigators, we find that some are not in good agreement. The disagreement may be attributed to two main causes: (1) The early investigators may not have excluded the observational data taken during primary eclipse. (2) An error may have been introduced by the use of normal points which were necessary during the era of desk computers. The first cause is more serious than the second, as we can see in table 9, where we have shown three sets of orbital elements, all derived from the recent Mount Wilson observations. The first set is derived from all observations including those during principal eclipse, the second set from all observations taken between 12.30 days (0.95 P) and 0.78 day (0.06 P), and the third set (the best one) from observations taken between 11.0 days (0.15 P) and 1.99 days (0.85 P). The difference in ω in these three sets is striking. However, for such a small eccentricity, the value of ω has very little physical significance.

From tables 8 and 9 (last column) it appears that except for the γ-velocity there is no systematic change with time in any of the orbital elements. However,

the change in γ which amounts to several km./sec. is too large to be attributed to a chance fluctuation. We do not know at present whether this is due to a difference in the wave lengths used by different investigators, or to the existence of a third body in the system. Moreover, a change with time in the surface activity of the star may also produce a variation in the γ-velocity.

DISCUSSION

The "classical" interpretation of the light curve and spectroscopic variations of Beta Lyrae have been discussed by several authors, most recently by Struve in the Russell Lecture of the American Astronomical Society for 1957. Certain aspects of this interpretation encounter difficulties which have not been removed by our observations. It may therefore be permissible to present here some tentative considerations which in view of their revolutionary character cannot yet be regarded as sufficiently well established to displace the earlier interpretation.

TABLE 9

ORBITAL ELEMENTS (AND MEAN ERRORS) OF β LYRAE, BERKELEY (1955)

Phase interval excluded....		12.30–0.78 days	11.0–1.99 days
γ	-16.0 ± 0.3 km./sec.	-17.5 ± 0.3 km./sec.	-15.0 ± 0.4 km./sec.
K	187.1 ± 0.5 km./sec.	186.1 ± 0.4 km./sec.	185.0 ± 0.3 km./sec.
e	0.007 ± 0.002	0.008 ± 0.002	0.017 ± 0.002
ω	$121° \pm 19°$	$330° \pm 14°$	$217° \pm 7°$
$T_0{}^*$	9.997 ± 0.005 days	9.964 ± 0.005 days	10.016 ± 0.008 days
T^*	1.425 ± 0.676 days	8.875 ± 0.495 days	4.884 ± 0.256 days
$a \sin i$	—	—	32.9×10^6 km
$f(\mathfrak{M})$	—	—	$8.5 \odot$
m.e. per plate	—	—	± 1.9 km./sec.
No. of plates	192	155	115

* After primary minimum, as determined from the formula used for computing the phases.

TABLE 10

Values of the Masses for Different Mass Ratios
$f(\mathfrak{M}) = 8.5\,\odot$

μ	$\mathfrak{M}_{Secondary}$	$\mathfrak{M}_{B8}$
2.0	76.5 $\odot$	153.0 $\odot$
1.5	53.1	79.6
1.0	34.0	34.0
0.6	21.7	12.4
0.5	19.1	9.6
0.4	16.7	6.7
0.3	14.4	4.3
0.2	12.2	2.4
0.16	11.4	1.8
0.1	10.3	1.0

The mass of a B8V, according to the mass luminosity relation, is about 4 $\odot$.

Table 10 lists a set of pairs of values for the masses of the two components, which are consistent with $f(\mathfrak{M}) = 8.5\,\odot$. The inclination was assumed to be $i = 90°$.

Kuiper [9] had reached the conclusion that the mass-ratio is of the order 1.5, implying masses of about 80 $\odot$ for the primary component and 50 $\odot$ for the secondary, and a luminosity of the order of -8 for the brighter star. However, Kuiper also remarked that the observational evidence favors an absolute magnitude of about -4.5 for this component, a value which is also suggested by the number of Balmer lines which are observed in the spectrum.

If we accept that the absolute magnitude of the principal component of Beta Lyrae is about -4.5 and that the mass-luminosity relation holds, the mass of the B8 star would be of the order of 13 $\odot$. The mass-function then yields a mass of about 22 $\odot$ for the secondary component. Consequently, we are led to the conclusion that the primary component of Beta Lyrae is less massive than the secondary.

Since a mass of 80 $\odot$ for the primary component seems to be too large for a star of a normal giant B8 spectrum and since the luminosity appears to be about -4.5 mag., one wonders whether we are not dealing with a system in which the brighter component is actually the less massive. This possibility is, in fact, supported by the early works of Belopolsky [23, 24] and of Curtiss [25]. In 1893 Belopolsky measured the radial velocities of the broad emissions at Hβ and derived a velocity curve opposite in phase to that from the absorption lines of the B8 component. Belopolsky's results were confirmed several years later by Curtiss, who derived a semi-amplitude of 75 km./sec. from the velocities of the Hβ emissions. From our material, Struve [11] estimated that the semi-amplitude from the broad emissions is of the order of 100 km./sec.

A velocity curve opposite in phase to that of the B8 component and with smaller amplitude could mean that either the material responsible for the emission moves with the secondary star or it is located between the two stars but on the side of the secondary component with respect to the center of gravity of the system. The presence of the emissions at all phases, even at 0.5 P, and their behavior during eclipse seem to rule out the possibility that the emitting gases are located between the two stars and suggest that they surround the secondary component and extend to a distance from its center larger than the radius of the B8 star. In the case of W Serpentis [26] the broad emissions in the spectrum shift in position with the same phase relationship as do the stellar absorption lines, and yield the same velocity amplitude as the latter. Moreover, in the Wolf-Rayet binaries, the motion of the Wolf-Rayet component seems to be given by the emission lines. It is therefore reasonable to believe that in Beta Lyrae the velocity amplitude from the broad emissions might roughly coincide with the velocity amplitude of the secondary star of the system. However, we must keep in mind that the matter which forms the streams and that which constitutes the expanding shell must contribute to the broad emissions; therefore, perhaps only a spectrophotometric study of the emission features would give us information about the relative contributions from each source.

The emission "peaks" which appear superimposed upon the broad emissions and are conspicuous during eclipse yield a semi-amplitude of the order of 30 km./sec. and a γ-velocity about 130 km./sec. more positive than that of the B8 component. The meaning of these results is not obvious, but an examination, especially of the region of λ3888 and the agreement between the measurements at He I 3888, Hγ and He I 4472, suggest that they might be real. The difference in γ then might be due to the fact that the emission "peaks" are parts of P Cygni-type profiles, the violet absorption edges being undetected because of the relatively faint continuous spectrum from the secondary component. Such a conclusion would imply that this component is an abnormal object: a very massive and underluminous star, which does not contribute appreciably to the light of the system and displays observable features only in the form of emissions. The satellite lines would still be interpreted as arising from the two streams mentioned earlier. The fact that there is a gaseous stream from the secondary star towards the primary must mean that the lobe of the inner contact surface around the secondary is filled with matter. Since the spectroscopic behavior of the system suggests that the radius of the secondary component is smaller than that of the primary, the matter filling the space between the secondary star and its lobe of the first critical equipotential surface must not be very opaque.

If we are willing to accept that the emission "peaks" and/or the broad emissions originate in matter which moves with the secondary star, we must also conclude that the mass of the secondary star is larger than that of the primary. The mass-ratio suggested by the ratio of the velocity amplitude of the B8 component to the velocity amplitude from the broad emissions implies (table 10) that the mass of the primary component must be about 7 ⊙ and that of the secondary about 17 ⊙. The mass-ratio obtained from the velocity amplitude of the emission "peaks" yields a mass of about 2 ⊙ for the primary and 11.5 ⊙ for the secondary.

A model of Beta Lyrae in which the mass of the secondary star is larger than that of the primary has certain advantages. For one thing, we would not need to attribute a very large mass to the B8 star. We could rather attribute to this component the mass which we would expect for a normal B8 giant.

According to the current ideas of stellar evolution, the primary component, being a B8 giant, would have expanded from its original size on the main sequence and would now be overluminous, in terms of the mass-luminosity relation. The existence of a gaseous stream from the star further suggests that it probably fills its lobe of the inner contact surface. On the other hand, the secondary component would be more massive than is consistent with the mass-luminosity relation. Since it is now more massive than the primary, it was almost certainly the more massive component when the system was formed and, therefore, would have evolved faster than the B8 star. Therefore, the former would be in a more advanced stage of evolution than the latter, being now underluminous for its mass.

One of the puzzles of Beta Lyrae which has intrigued astronomers for many years is the slow velocity of rotation of the B8 component [3]. While synchronization between axial rotation and orbital revolution required in the "classical" model of Beta Lyrae a velocity of rotation greater than 200 km./sec., the profiles of the absorption lines show that at present V_{rot} = 40–50 km./sec. [27, 28, 29]. Table 11 shows that if we allow the mass-ratio μ to be less than one, the discrepancy in the velocity of rotation becomes less serious [11, 4]. The remaining difference could be explained in terms of evolution [30, 31], since the

TABLE 11

MINIMUM RADIUS AND SYNCHRONIZED ROTATIONAL VELOCITY OF THE B8 COMPONENT FOR DIFFERENT MASS-RATIOS, ASSUMING THAT THE STAR FILLS ITS LOBE OF THE INNER CONTACT SURFACE

μ	Minimum R	Synchronized V_{rot}
0.4	20 ⊙	110 km./sec.
0.3	14	80
0.15	12	68

The radius of a B8V star is of the order of 3.5 ⊙.

B8 star has expanded from its original size and now fills its lobe of the corresponding inner contact surface.

Since we cannot be certain whether the motion of the secondary star is shown by the broad emissions or the narrow "peaks" superimposed upon them, we can do little more than suggest that the mass of the secondary may be considerably larger than that of the primary, and that the secondary star certainly and the primary quite probably depart from the conventional mass-luminosity relation. This conclusion, as far as the masses are concerned, is not new: it was recently discussed by S. Gaposchkin [32] (though in a somewhat different manner) and it was mentioned by Struve [11, 4].

The light curve of Beta Lyrae can be satisfied by a variety of models in which the primary component is either larger or smaller than the secondary. If we impose the condition that the primary star be of smaller mass than the secondary and that it fills its lobe of the inner contact surface, the body which is in front during the principal eclipse will be larger than the B8 star. Since the spectroscopic observations indicate that the size of the eclipsing star must be smaller, a new photometric solution would have to consider the case of a small secondary star surrounded by a very extended, tenuous envelope which is essentially transparent to the continuous spectrum of the B8 star with its absorption lines.

The observations of Beta Lyrae show that the time of primary minimum falls somewhat later than was predicted by the expression which we have used to derive the phases. It is probably safe to derive the present value for the correction to the predicted times of minima by computing the time of superior conjunction from the orbital elements in table 9 (last column). The result is 0.379 day or 0.0293 P later than the predicted times of minima. Allowing for this correction and considering the information contained in Struve's 1941 paper [6], we conclude that the red satellites are present at least in the phase interval 0.92 P −0.98 P and that the violet satellites are present in the phase interval 0.001 P −0.06 P. This implies that the stream responsible for the red satellite lines is seen projected upon a region of the B8 star which is located closer to the radius vector joining the two components than the region upon which the stream of the violet satellites is seen projected. The receding (red) stream apparently originates in the neighborhood of the Lagrangian point L_1 of the restricted three bodies problem. Since the period of rotation of the primary star is larger than the orbital period [3], it is tempting to locate the origin of the stream which produces the violet satellite lines near the tip of the primary star in a region slightly displaced in the direction opposite to that of the rotation.

REFERENCES

1. STRUVE, O.   Pub. Astr. Soc. Pacific **70**: 5–40, 1958.
2. STRUVE, O.   Am. Jour. Phys. **9**: 63–80, 1941.
3. STRUVE, O.   Pub. Astr. Soc. Pacific **64**: 180–184, 1952.
4. STRUVE, O.   Sky and Telescope **16**: 418–422, 1957.
5. KOPAL, Z.   Astrophys. Jour. **93**: 92–103, 1941.
6. STRUVE, O.   Astrophys. Jour. **93**: 104–117, 1941.
7. GILL, J. R.   Astrophys. Jour. **93**: 118–127, 1941.
8. GREENSTEIN, J. L., AND T. L. PAGE.   Astrophys. Jour. **93**: 128–132, 1941.
9. KUIPER, G. P.   Astrophys. Jour. **93**: 133–177, 1941.
10. SHERMAN, F.   Astrophys. Jour. **94**: 368–374, 1941.
11. STRUVE, O.   International Astr. Union Symposium on Non-stable stars (ed.: G. H. Herbig; Cambridge University Press), 93–107, 1957.
12. PRAGER, R.   Kl. Veröff. Berlin-Babelsberg **3** (No. 10): 125, 1931.
13. SAIDOV, K.   Astr. Circ. Academy of Sci. Soviet Union, No. 158: 12, 1955.
14. SAHADE, J., AND C. U. CESCO.   Astrophys. Jour. **101**: 235–239, 1945.
15. SAHADE, J., AND O. STRUVE.   Astrophys. Jour. **102**: 480–491, 1945.
16. STRUVE, O., AND S.-S. HUANG.   Handbuch der Physik (ed.: S. Flügge; Springer-Verlag, Heidelberg), **50**: 243–273, 1957.
17. LEHMANN-FILHÉS, R.   Astron. Nachr. **136**: 17–30, 1894.
18. SCHLESINGER, F.   Pub. Allegheny Obs. **1**: 33–44, 1908.
19. SCHLESINGER, F., AND S. UDICK.   Pub. Allegheny Obs. **2**: 155–190, 1912.
20. LUYTEN, W. J.   Pub. Astr. Obs. Univ. Minnesota **2**: 53–71, 1936.
21. STERNE, T. E.   Proc. Nat. Academy Sci. **27**: 175–181, 1941 (Harvard Reprint No. 222).
22. ROSSITER, R. A.   Astrophys. Jour. **60**: 15–21, 1924.
23. BELOPOLSKY, A.   Mem. della Soc. Spett. Italiani **22**: 101–111, 1893; **26**: 135–143, 1897; Astrophys. Jour. **6**: 328–337, 1897.
24. TIKHOFF, G.   Mem. della Soc. Spett. Italiani **26**: 107–112, 1897.
25. CURTISS, H. R.   Pub. Allegheny Obs. **2**: 73–120, 1912.
26. SAHADE, J., AND O. STRUVE.   Astrophys. Jour. **126**: 87–98, 1957.
27. WARES, G.   Quoted in reference 3.
28. BÖHM-VITENSE, E.   Astrophys. Jour. **120**: 271–273, 1954.
29. MITCHELL, R. I.   Astrophys. Jour. **120**: 274–277, 1954.
30. SANDAGE, A. R.   Astrophys. Jour. **122**: 263–270, 1955.
31. KOPAL, Z.   Annales d'Astrophysique **19**: 298–335, 1956.
32. GAPOSCHKIN, S.   Zeitschr. f. Astrophysik **39**: 133–136, 1956.

ATLAS

Following are thirty photographic enlargements of spectrograms in the region λλ3680–4580.

Table 2 in the text gives the correspondence between the phases and the calendar dates.

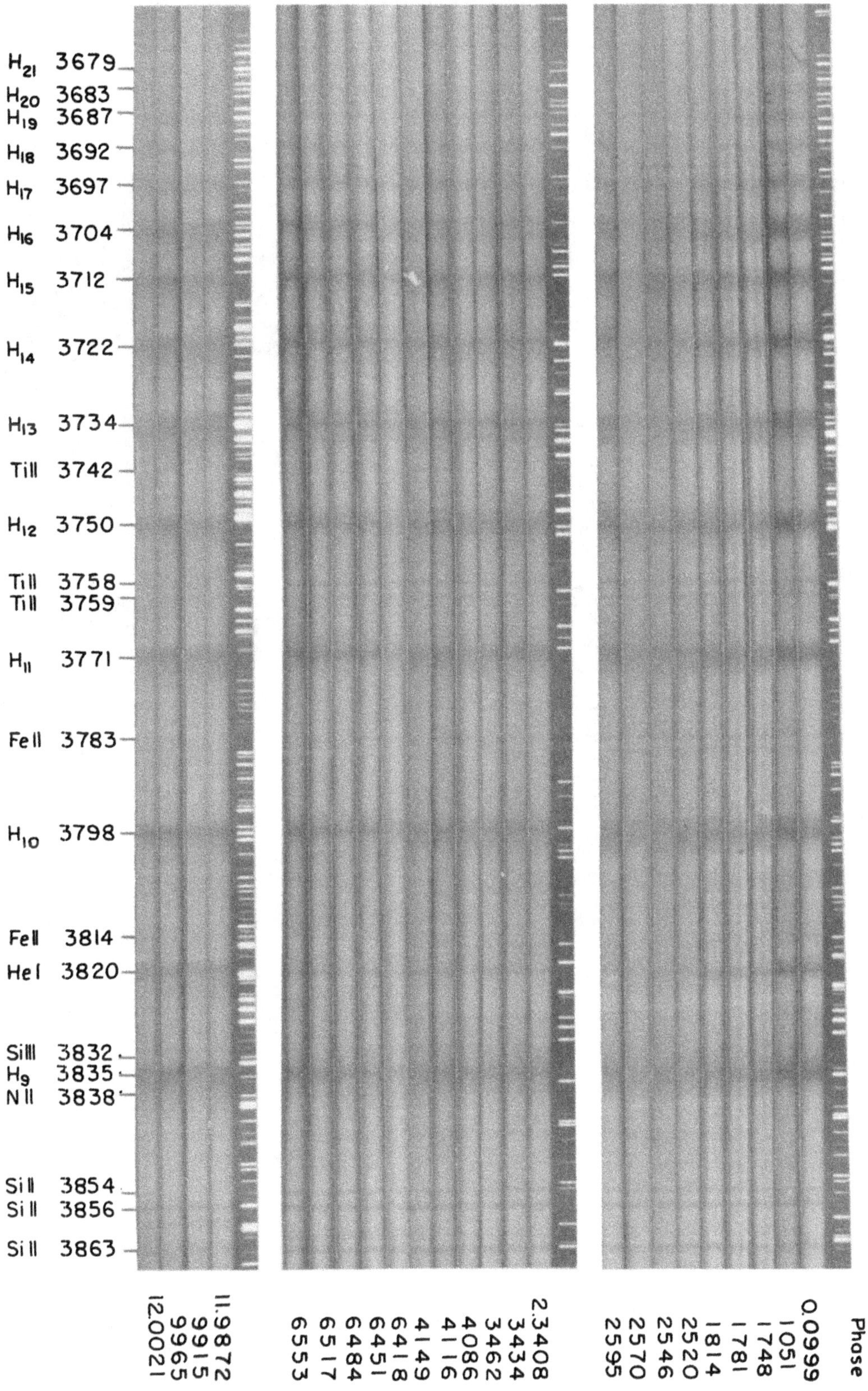

H$_{21}$ 3679
H$_{20}$ 3683
H$_{19}$ 3687
H$_{18}$ 3692
H$_{17}$ 3697
H$_{16}$ 3704
H$_{15}$ 3712
H$_{14}$ 3722
H$_{13}$ 3734
TiII 3742
H$_{12}$ 3750
TiII 3758
TiII 3759
H$_{11}$ 3771
FeII 3783
H$_{10}$ 3798
FeII 3814
HeI 3820
SiIII 3832
H$_9$ 3835
NII 3838
SiII 3854
SiII 3856
SiII 3863

11.9872
9915
9965
12.0021

2.3408
3434
3462
4086
4116
4149
6418
6451
6484
6517
6553

0.0999
1051
1748
1781
1814
2520
2546
2570
2595

Phase

35

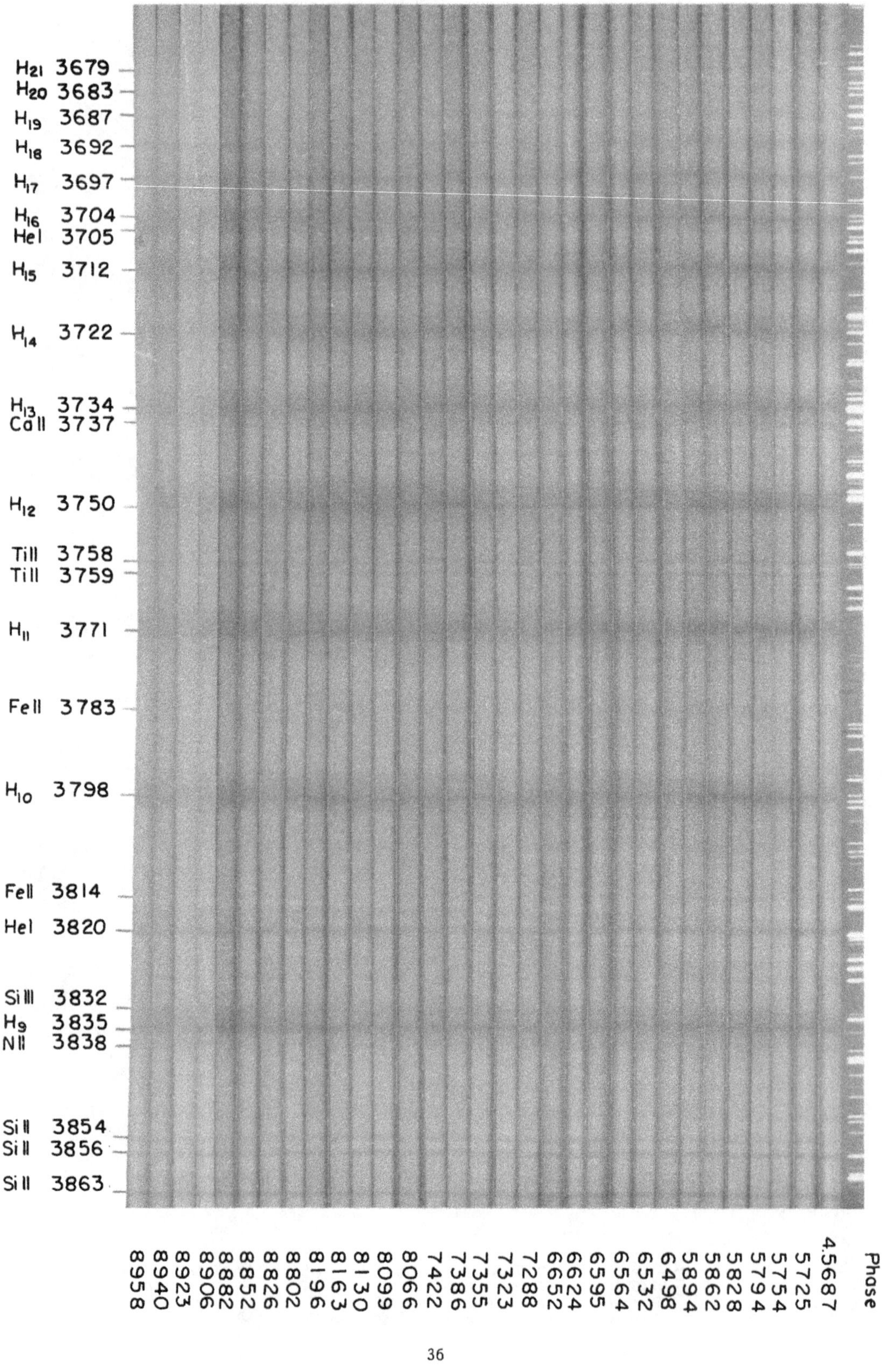

H21 3679
H20 3683
H19 3687
H18 3692
H17 3697
H16 3704
HeI 3705
H15 3712
H14 3722
H13 3734
CaII 3737
H12 3750
TiII 3758
TiII 3759
H11 3771
FeII 3783
H10 3798
FeII 3814
HeI 3820
SiIII 3832
H9 3835
NII 3838
SiII 3854
SiII 3856
SiII 3863

8958 8940 8923 8906 8882 8852 8826 8802 8163 8196 8130 8099 8066 7422 7386 7355 7323 7288 6652 6624 6595 6564 6532 6498 5894 5862 5828 5794 5754 5725 4.5687

Phase

36

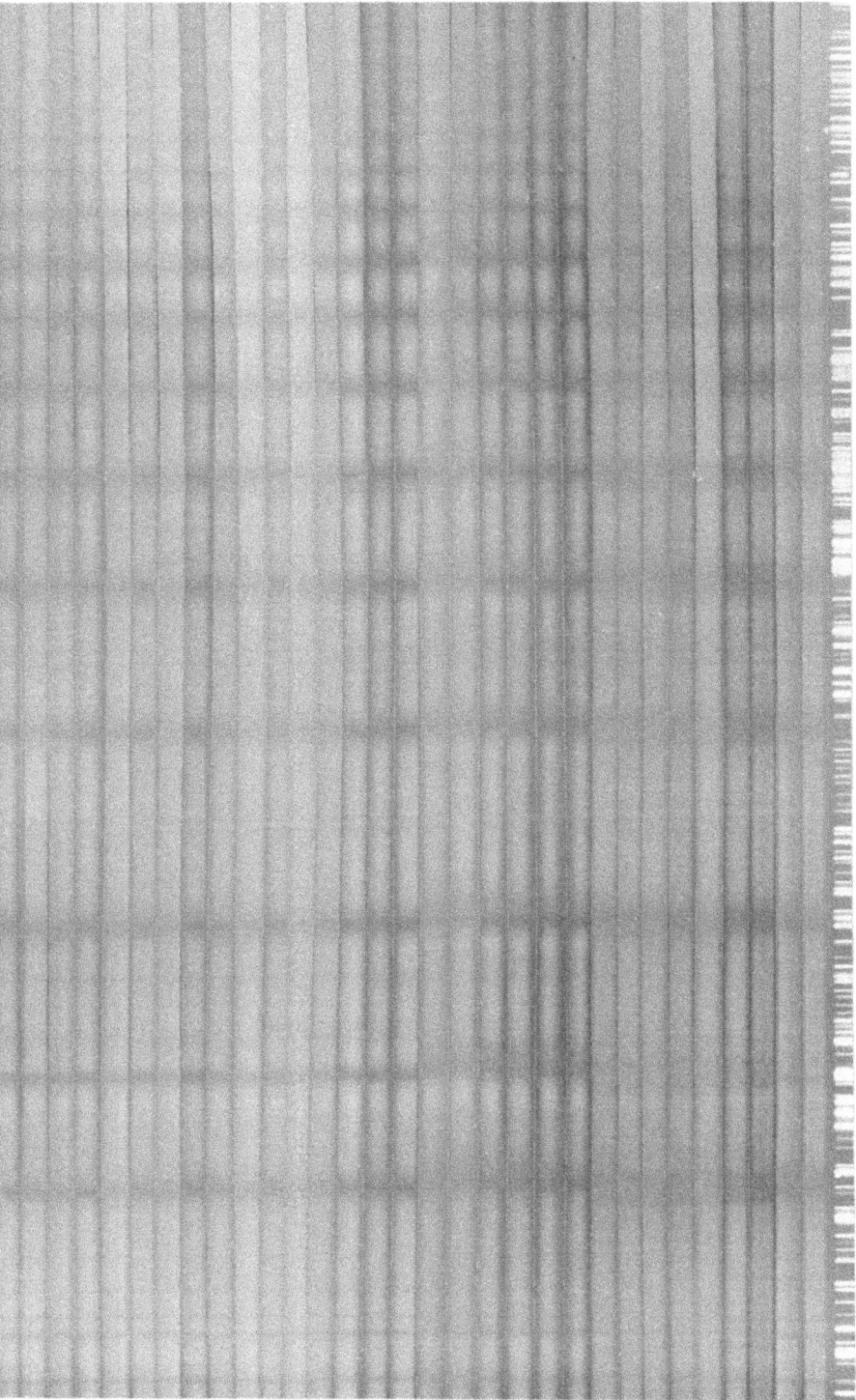

Line	λ
H$_{21}$	3679
H$_{20}$	3683
H$_{19}$	3687
H$_{18}$	3692
H$_{17}$	3697
H$_{16}$	3704
H$_{15}$	3712
H$_{14}$	3722
H$_{13}$	3734
Ca	3737
Ti II	3742
H$_{12}$	3750
Ti II	3758
Ti II	3759
H$_{11}$	3771
Fe II	3783
H$_{10}$	3798
Fe II	3814
He I	3820
H$_{9}$	3835
Si II	3854
Si II	3856
Si II	3863

Phase

48974
8991
9552
9586
9614
9642
9675
9712
9753
5.03365
0397
0431
0467
0495
0518
1092
1114
1133
1147
1159
1170
1182
1193
1204
1215
1232
1249
1265
1282
1299
1314

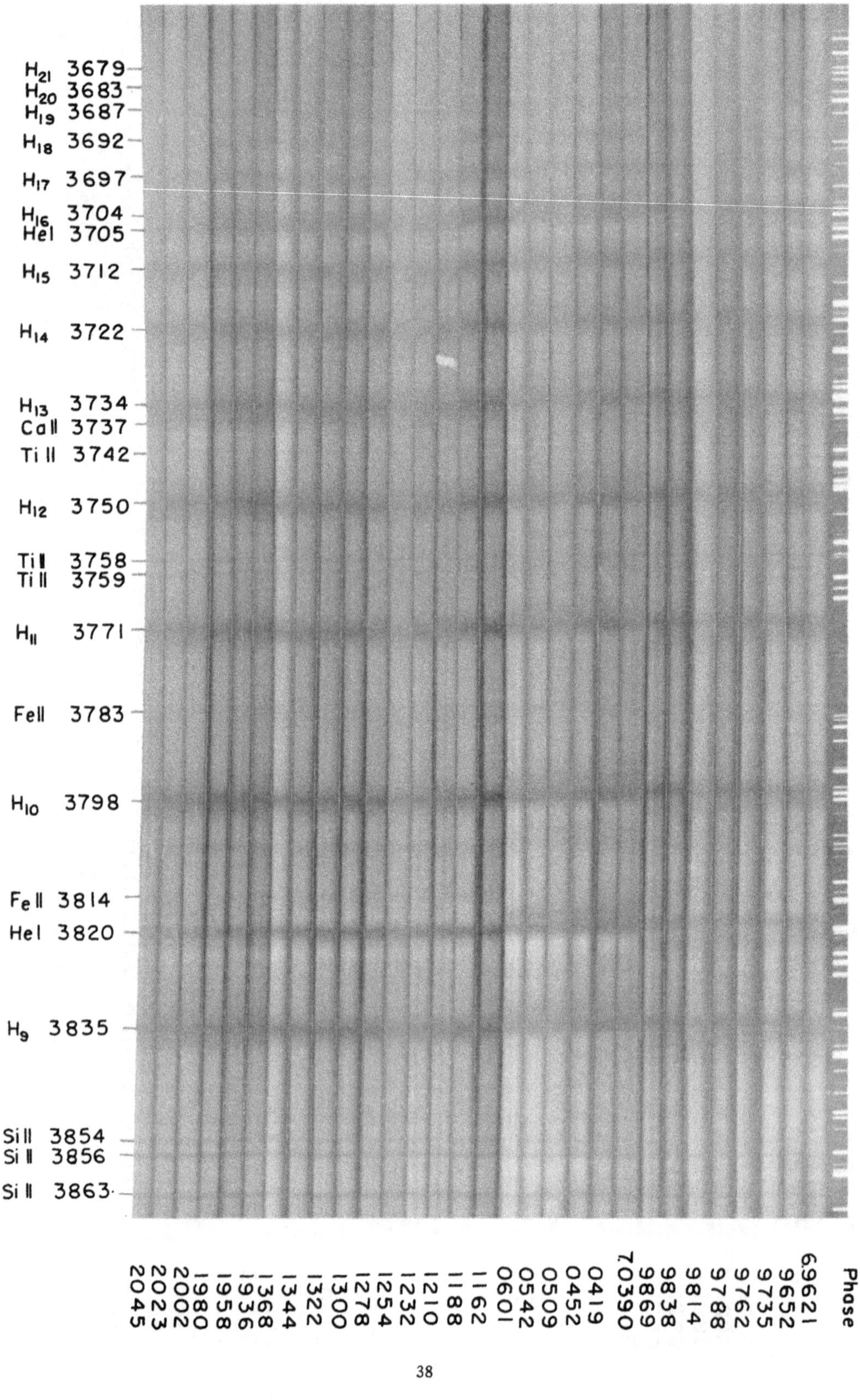

H₂₁	3679	
H₂₀	3683	
H₁₉	3687	
H₁₈	3692	
H₁₇	3697	
H₁₆	3704	
HeI	3705	
H₁₅	3712	
H₁₄	3722	
H₁₃	3734	
CaII	3737	
TiII	3742	
H₁₂	3750	
TiII	3758	
TiII	3759	
H₁₁	3771	
FeII	3783	
H₁₀	3798	
FeII	3814	
HeI	3820	
H₉	3835	
SiII	3854	
SiII	3856	
SiII	3863	

Labels at left of spectrogram:

H_{21} 3679
H_{20} 3683
H_{19} 3687
H_{18} 3692
H_{17} 3697
H_{16} 3704
HeI 3705
H_{15} 3712
H_{14} 3722
H_{13} 3734
CaII 3737
TiII 3742
H_{12} 3750
TiII 3758
TiII 3759
H_{11} 3771
FeII 3783
H_{10} 3798
Fe_{II} 3814
HeI 3820
H_{9} 3835
SiII 3854
SiII 3856
SiII 3863

Phase values (bottom axis):

2045
2023
2002
1980
1958
1936
1368
1344
1322
1300
1278
1254
1232
1210
1188
1162
0601
0542
0509
0452
0419
70390
9869
9838
9814
9788
9762
9735
9652
6.9621

Phase

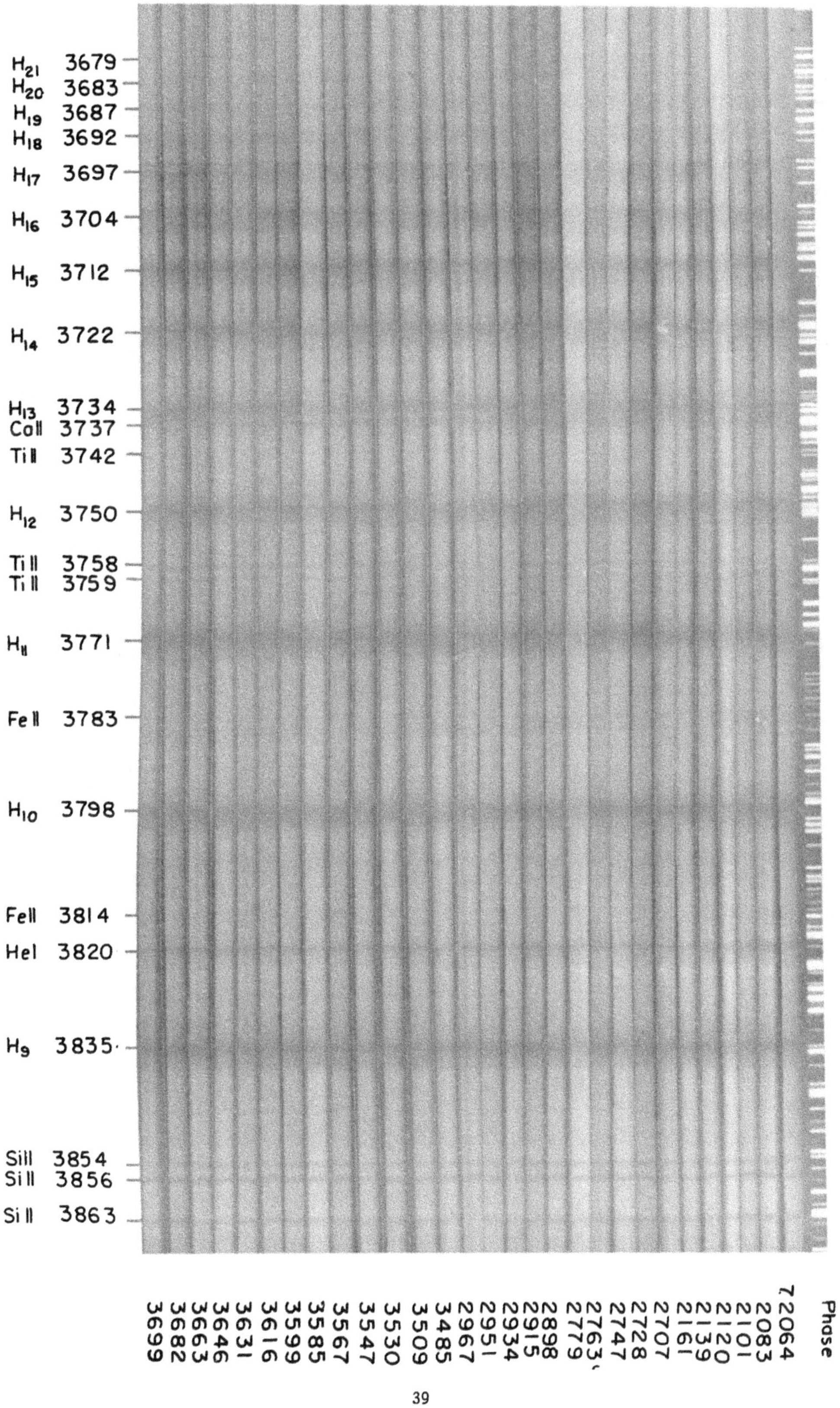

		Phase
H₂₁ 3679		7 2064
H₂₀ 3683		2083
H₁₉ 3687		2101
H₁₈ 3692		2120
H₁₇ 3697		2139
H₁₆ 3704		2161
H₁₅ 3712		2707
H₁₄ 3722		2728
H₁₃ 3734		2747
Co II 3737		2763
Ti II 3742		2779
H₁₂ 3750		2898
Ti II 3758		2915
Ti II 3759		2934
H₁₁ 3771		2951
Fe II 3783		2967
H₁₀ 3798		3485
Fe II 3814		3509
He I 3820		3530
H₉ 3835		3547
Si II 3854		3567
Si II 3856		3585
Si II 3863		3599

Labels along the top axis (wavelengths):
H₂₁ 3679, H₂₀ 3683, H₁₉ 3687, H₁₈ 3692, H₁₇ 3697, H₁₆ 3704, H₁₅ 3712, H₁₄ 3722, H₁₃ 3734, Co II 3737, Ti II 3742, H₁₂ 3750, Ti II 3758, Ti II 3759, H₁₁ 3771, Fe II 3783, H₁₀ 3798, Fe II 3814, He I 3820, H₉ 3835, Si II 3854, Si II 3856, Si II 3863

Bottom axis values:
3699, 3682, 3663, 3646, 3631, 3616, 3599, 3585, 3567, 3547, 3530, 3509, 3485, 2967, 2951, 2934, 2915, 2898, 2779, 2763, 2747, 2728, 2707, 2161, 2139, 2120, 2101, 2083, 7 2064

Phase

H₂₁ 3679
H₂₀ 3683
H₁₉ 3687
H₁₈ 3692
H₁₇ 3697
H₁₆ 3704
HeI 3705
H₁₅ 3712
H₁₄ 3722
H₁₃ 3734
CaII 3737
TiII 3742
FeII 3748
H₁₂ 3750
TiII 3758
TiII 3759
H₁₁ 3771
FeII 3783
H₁₀ 3798
FeII 3814
HeI 3820
SiIII 3832
H₉ 3835
NII 3838
SiII 3854
SiII 3856
SiII 3863

5274 5255 5234 5210 5190 5172 5151 5130 5108 5086 5060 5031 5000 4478 4459 4440 4425 4408 4392 4373 4357 4341 4324 4308 4292 4275 4254 3735 .73717

Phase

40

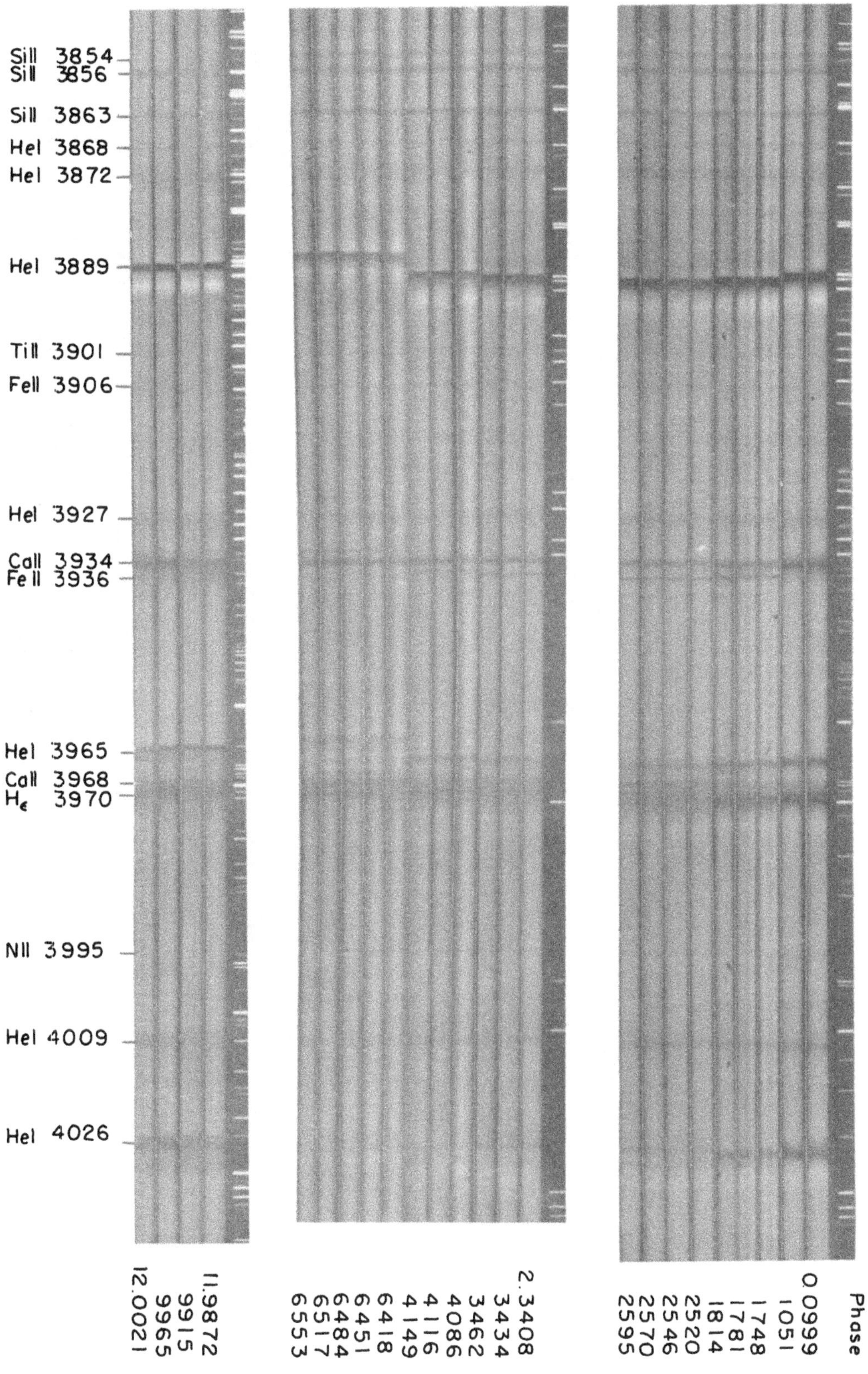

SiII 3854
SiII 3856

SiII 3863

HeI 3868

HeI 3872

HeI 3889

TiII 3901

FeII 3906

HeI 3927

CaII 3934
FeII 3936

HeI 3965

CaII 3968
H$_\epsilon$ 3970

NII 3995

HeI 4009

HeI 4026

Phase

12.0021
9965
9915
11.9872

6553
6517
6484
6451
6418
4149
4116
4086
3462
3434
2.3408

2595
2570
2546
2520
1814
1781
1748
1051
0.0999

41

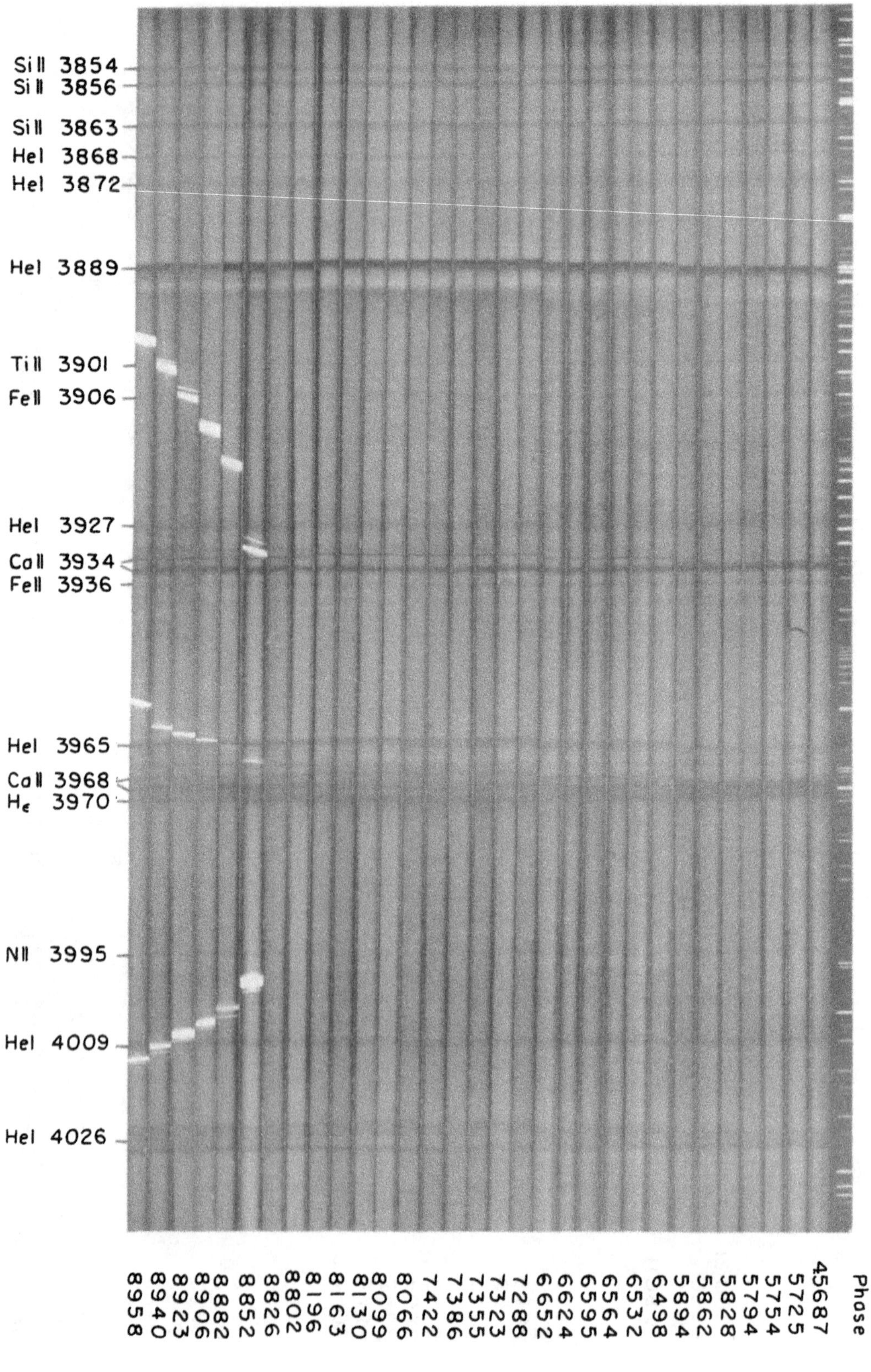

SiII 3854
SiII 3856
SiII 3863
HeI 3868
HeI 3872
HeI 3889
TiII 3901
FeII 3906
HeI 3927
CaII 3934
FeII 3936
HeI 3965
CaII 3968
H$_\epsilon$ 3970
NII 3995
HeI 4009
HeI 4026

Phase

8958 8940 8923 8906 8882 8852 8826 8802 8196 8163 8130 8099 8066 7422 7386 7355 7323 7288 7252 6652 6624 6595 6564 6532 6498 5894 5862 5828 5794 5754 5725 45687

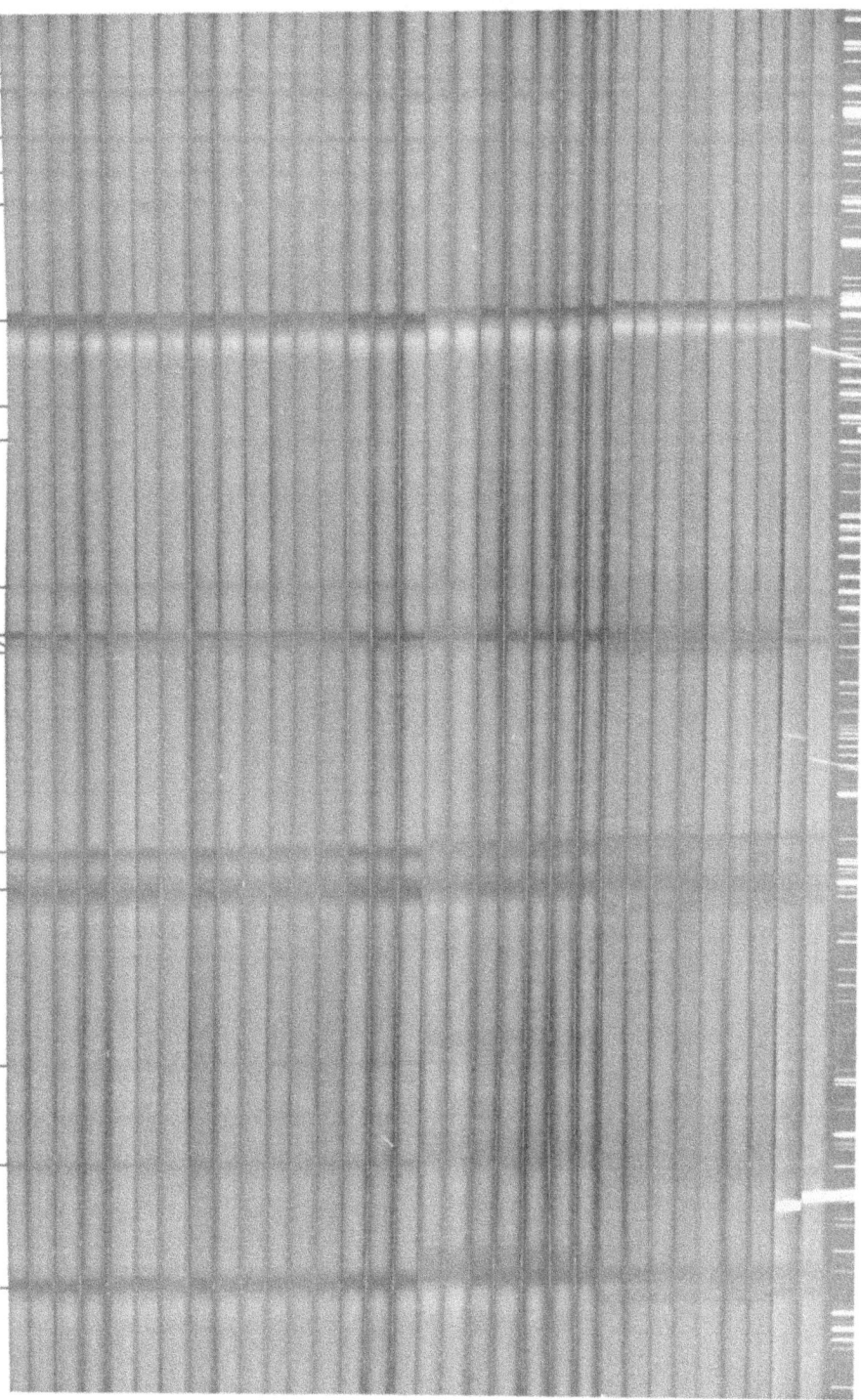

SiII 3854
SiII 3856

SiII 3863

HeI 3868

HeI 3872

HeI 3889

TiII 3901

FeII 3906

HeI 3927

CaII 3934
FeII 3936

HeI 3965

CaII, H$_\epsilon$

NII 3995

HeI 4009

HeI 4026

Phase

4.8974
8991
9552
9586
9614
9642
9675
9712
9753
5.0314
0365
0397
0431
0467
0495
0518
1092
1114
1133
1147
1159
1170
1182
1193
1204
1215
1232
1249
1265
1282
1299
1314

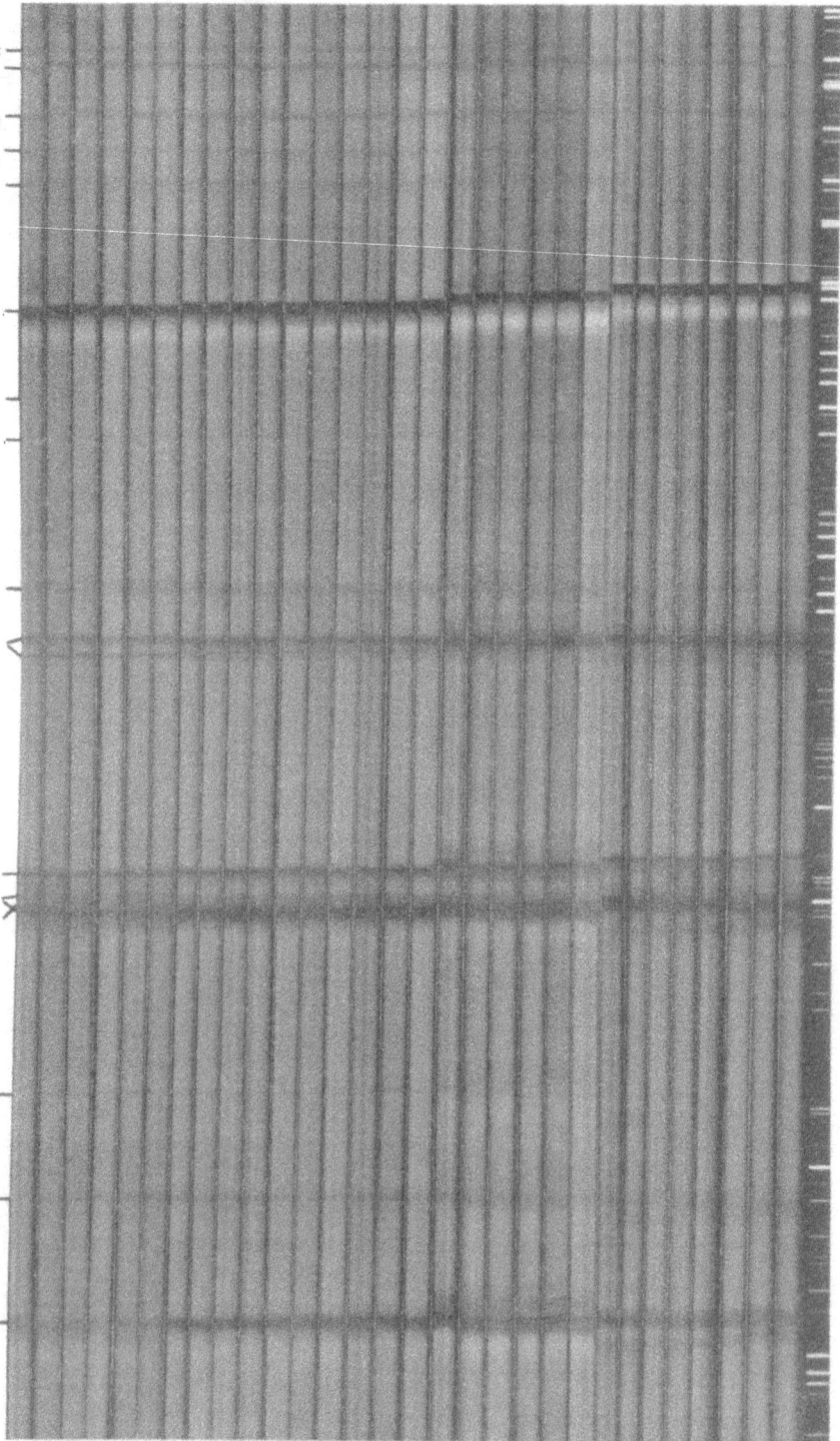

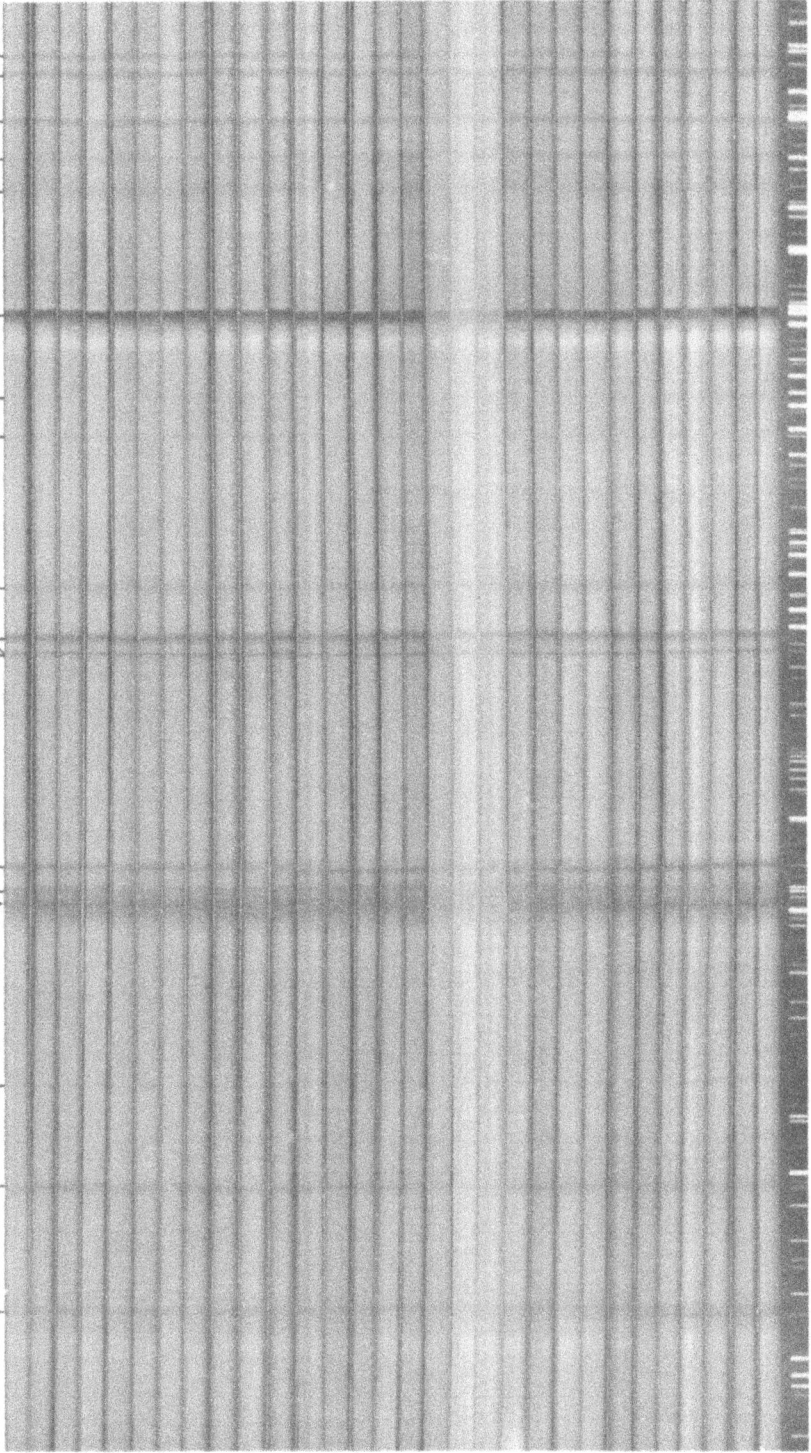

Sill 3854 —
Sill 3856 —

Sill 3863 —

Hel 3868 —

Hel 3872 —

Hel 3889 —

Till 3901 —

Fell 3906 —

Hel 3927 —

Ca 3934 —
Fell 3936 —

Hel 3965 —
Call 3968 —
H$_\epsilon$ 3970 —

NII 3995 —

Hel 4009 —

Hel 4026 —

Phase
7.2064
2083
2101
2120
2139
2161
2707
2728
2747
2763
2779
2797
2814
2831
2898
2915
2934
2951
2967
2951
3485
3509
3530
3547
3567
3585
3599
3616
3631
3646
3663

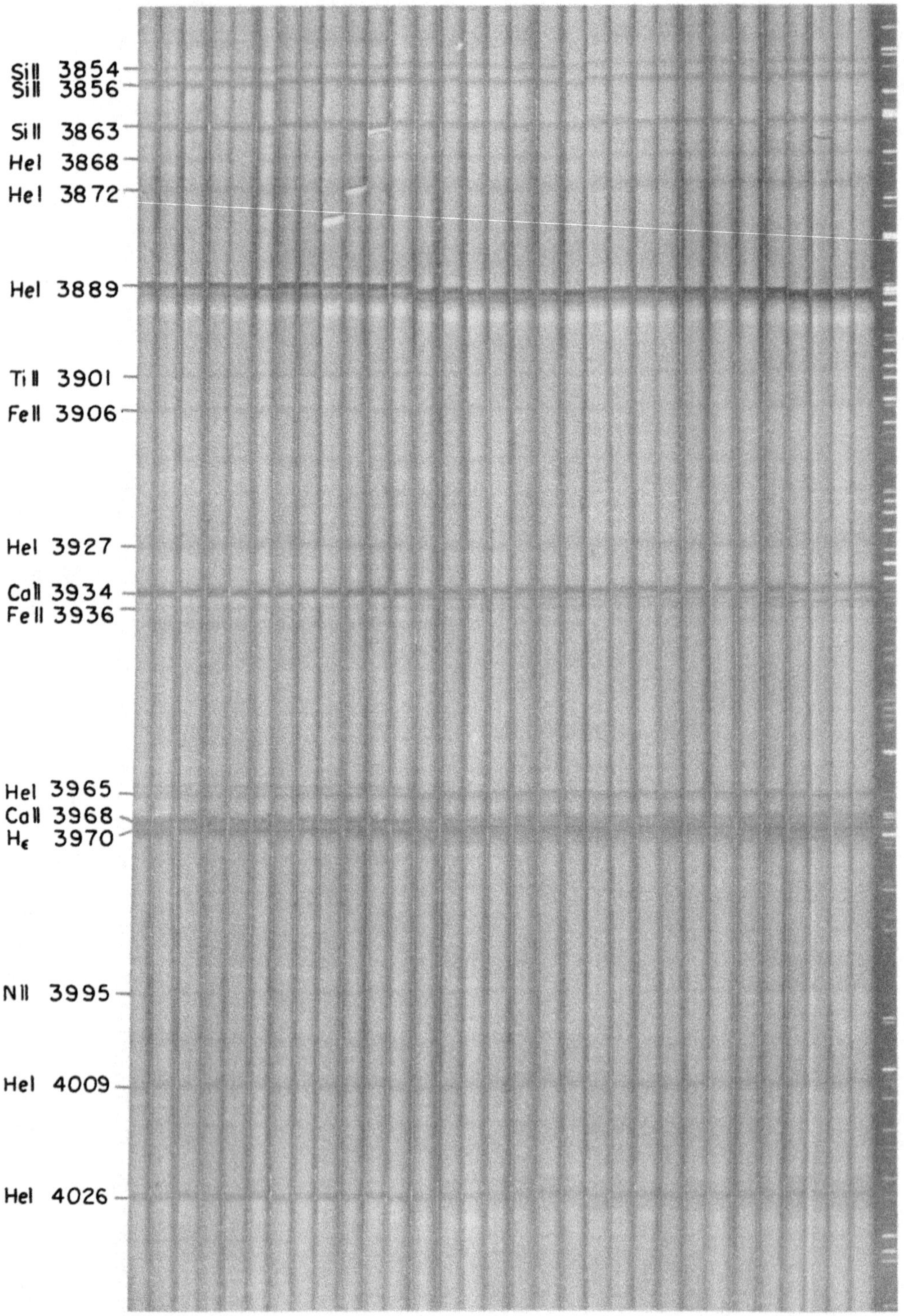

SiII 3854
SiII 3856

SiII 3863
HeI 3868
HeI 3872

HeI 3889

TiII 3901
FeII 3906

HeI 3927

CaII 3934
FeII 3936

HeI 3965
CaII 3968
H$_\epsilon$ 3970

NII 3995

HeI 4009

HeI 4026

5274
5255
5234
5210
5190
5172
5151
5130
5108
5086
5060
5031
5000
4500
4478
4459
4440
4425
4408
4392
4373
4357
4341
4324
4308
4292
4275
4254
3735
3717
3699
73682

Phase

46

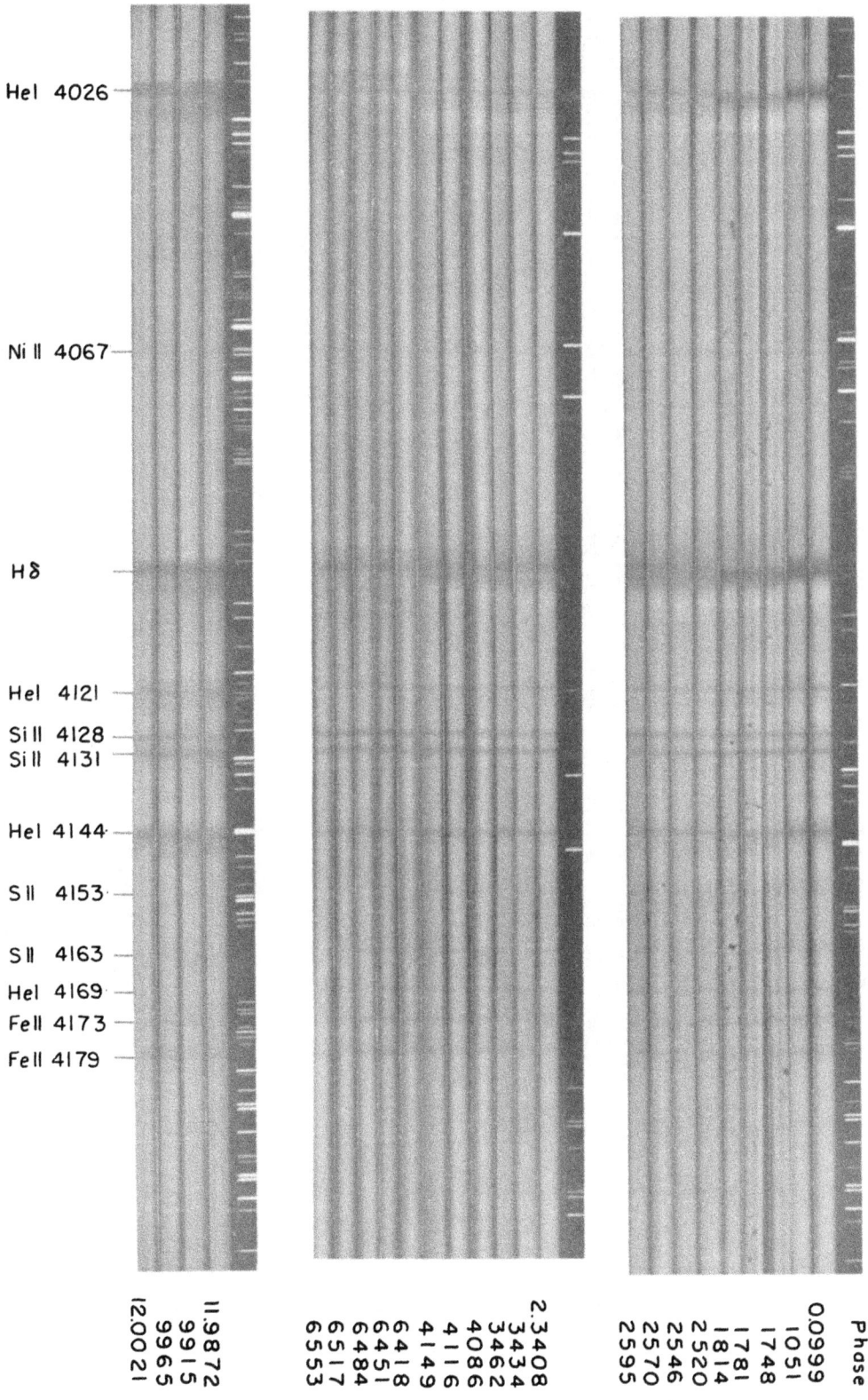

HeI 4026

NiII 4067

Hδ

HeI 4121
SiII 4128
SiII 4131

HeI 4144

SII 4153

SII 4163

HeI 4169
FeII 4173

FeII 4179

Phase

11.9872
9915
9965
12.0021

2.3408
3434
3462
4086
4116
4149
6418
6451
6484
6517
6553

0.0999
1051
1748
1781
1814
2520
2546
2570
2595

47

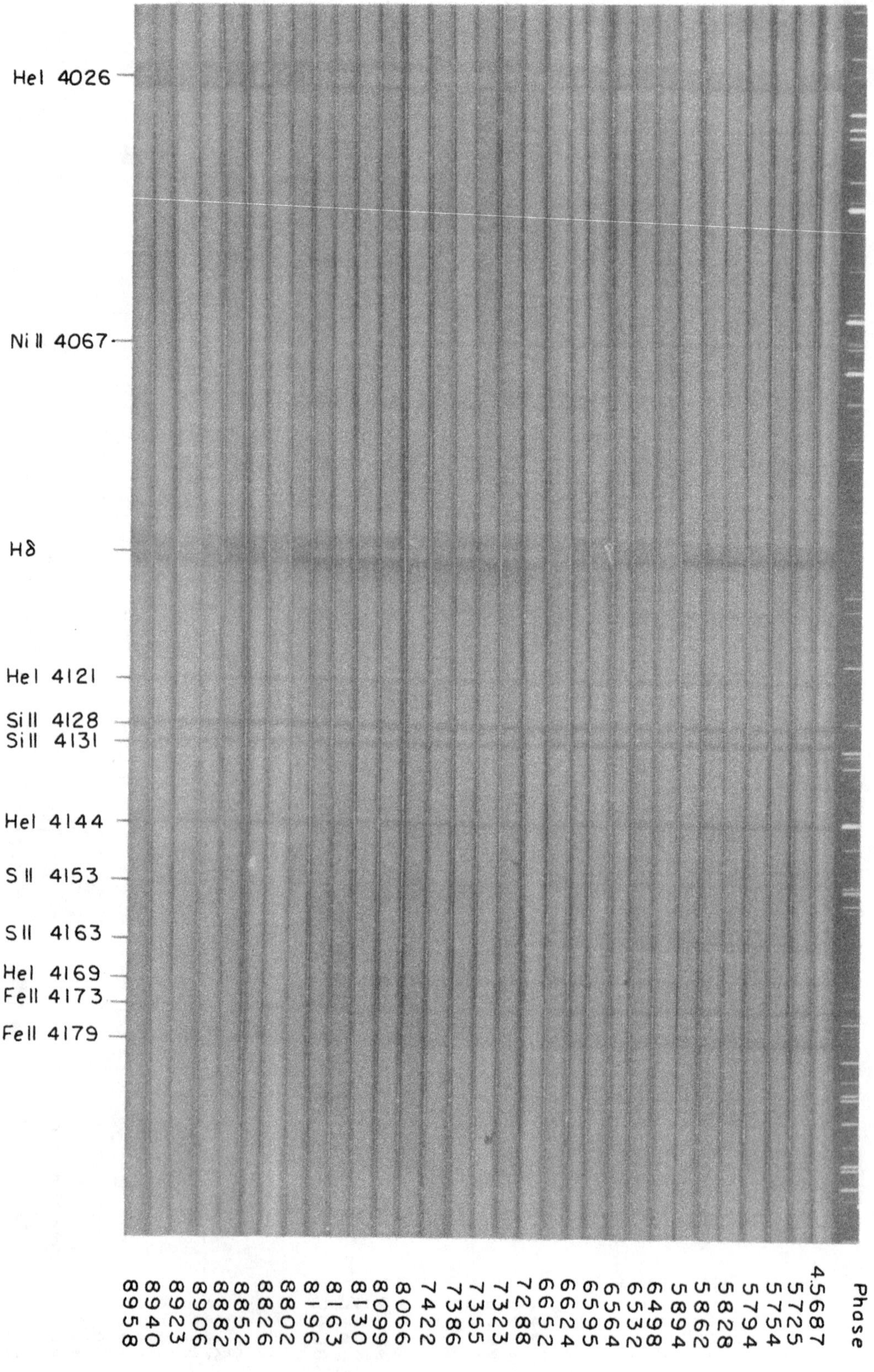

Hel 4026 —

Ni ll 4067 —

Hδ —

Hel 4121 —
Si ll 4128 —
Si ll 4131 —

Hel 4144 —

S ll 4153 —

S ll 4163 —
Hel 4169 —
Fell 4173 —

Fell 4179 —

8958 8940 8923 8906 8882 8852 8826 8802 8196 8163 8130 8099 8066 7422 7386 7355 7323 7288 7252 6652 6624 6595 6564 6532 6498 5894 5862 5828 5794 5754 5725 4.5687

Phase

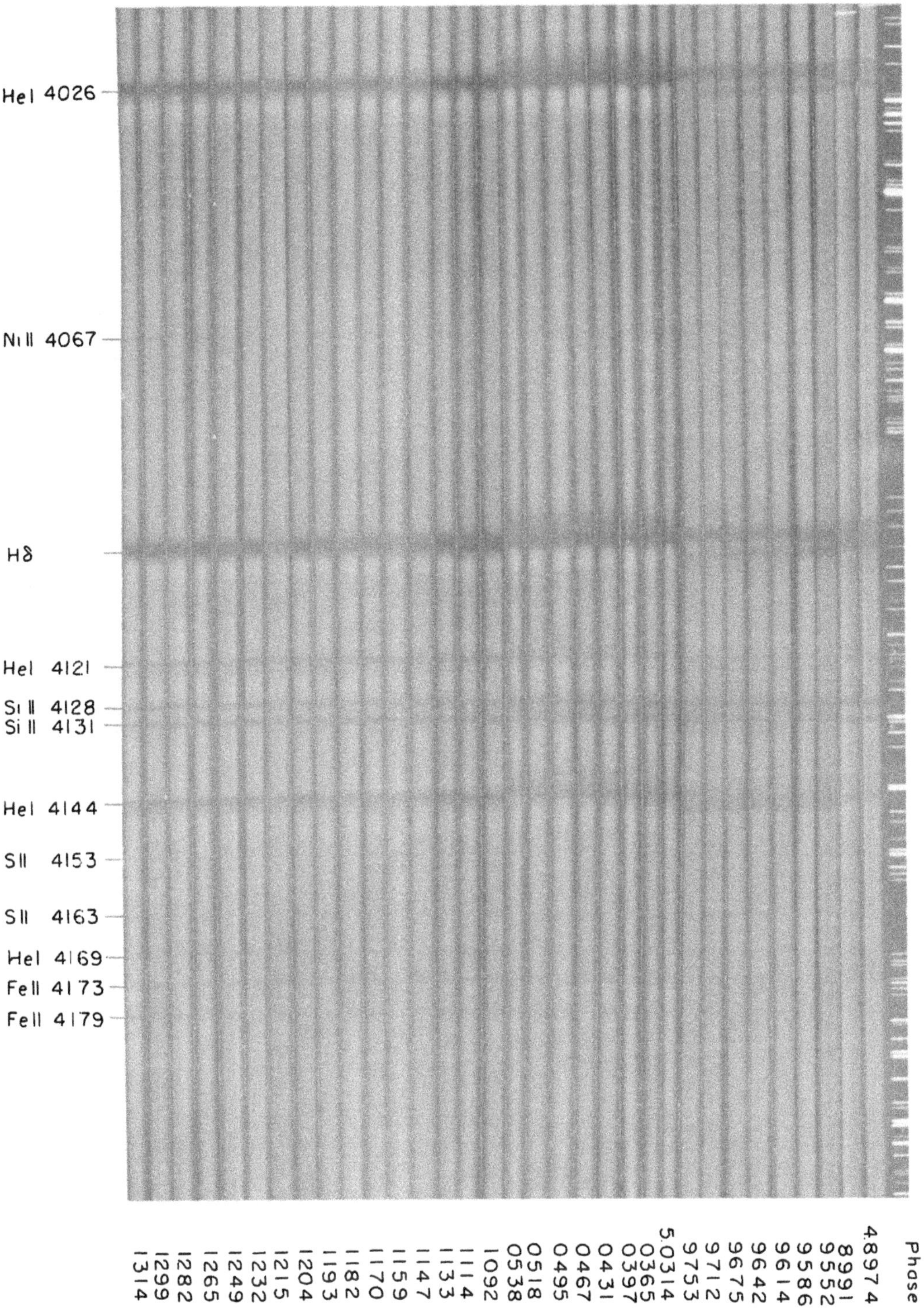

HeI 4026

NiII 4067

Hδ

HeI 4121
SiII 4128
SiII 4131

HeI 4144

SII 4153

SII 4163

HeI 4169
FeII 4173
FeII 4179

Phase

4.8974
.8991
.9552
.9586
.9614
.9642
.9675
.9712
.9753
5.0314
.0365
.0397
.0431
.0467
.0495
.0518
.0538
.1092
.1114
.1133
.1147
.1159
.1170
.1182
.1193
.1204
.1215
.1232
.1249
.1265
.1282
.1299
.1314

49

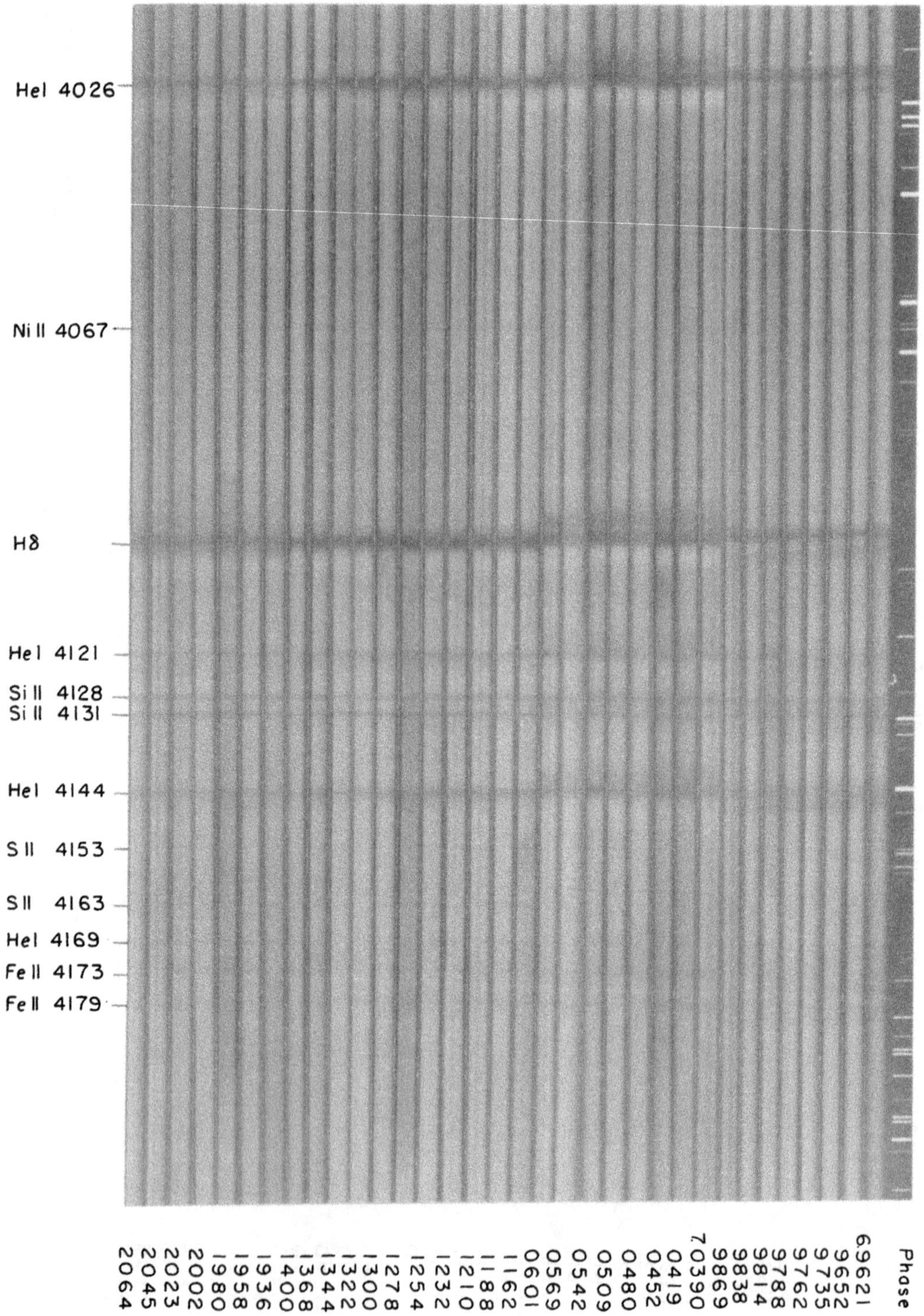
He I 4026
Ni II 4067
H8
He I 4121
Si II 4128
Si II 4131
He I 4144
S II 4153
S II 4163
He I 4169
Fe II 4173
Fe II 4179
Phase
6.9621
9652
9735
9762
9788
9814
9838
9869
7.0390
0419
0452
0480
0509
0542
0569
0601
1162
1188
1210
1232
1254
1278
1300
1322
1344
1368
1400
1936
1958
1980
2002
2023
2045
2064

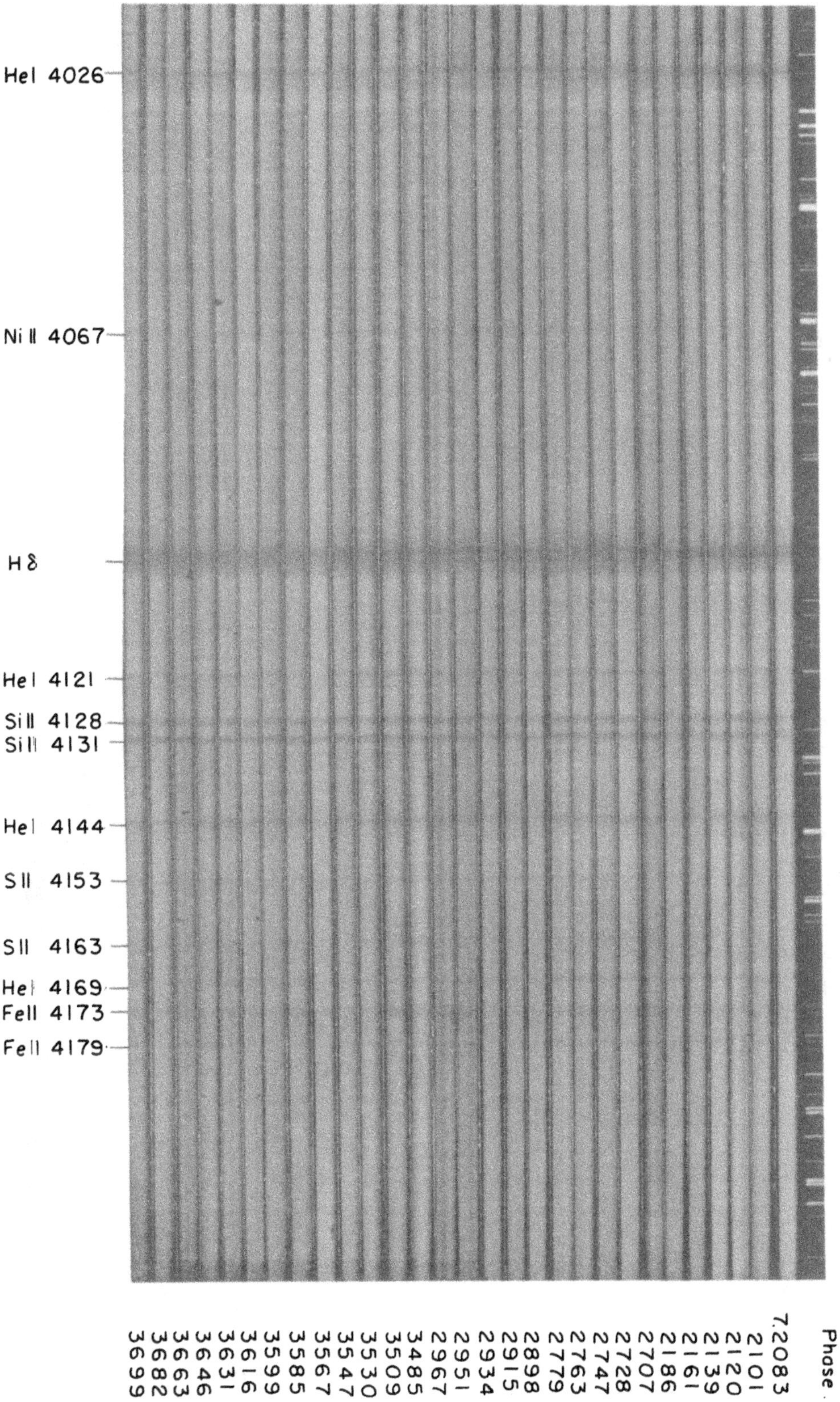

HeI 4026

NiII 4067

Hδ

HeI 4121

SiII 4128
SiII 4131

HeI 4144

SII 4153

SII 4163

HeI 4169
FeII 4173

FeII 4179

Phase

7.2083
2101
2120
2139
2161
2186
2186
2707
2728
2747
2763
2779
2898
2915
2934
2951
2967
2934
3485
3509
3530
3547
3567
3585
3599
3616
3631
3646
3663
3682
3699

51

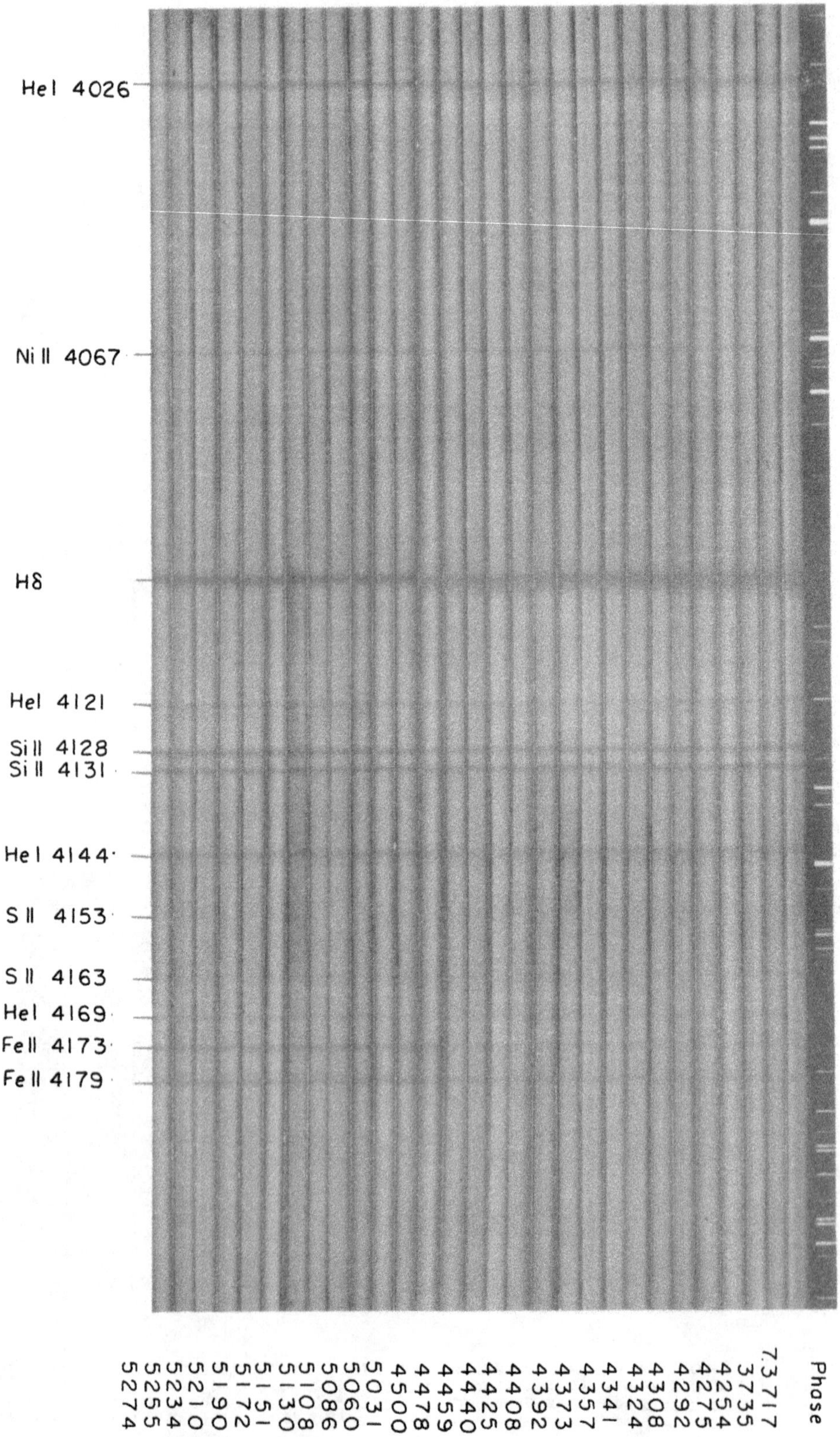

HeI 4026

NiII 4067

Hδ

HeI 4121

SiII 4128
SiII 4131

HeI 4144

SII 4153

SII 4163

HeI 4169

FeII 4173

FeII 4179

Phase
7.3 717
3735
42254
42275
42292
43308
43324
43341
43357
43373
43392
43408
4408
4425
4440
4459
4478
4500
5031
5060
5086
5108
5130
5151
5172
5190
5210
5234
5255
5274

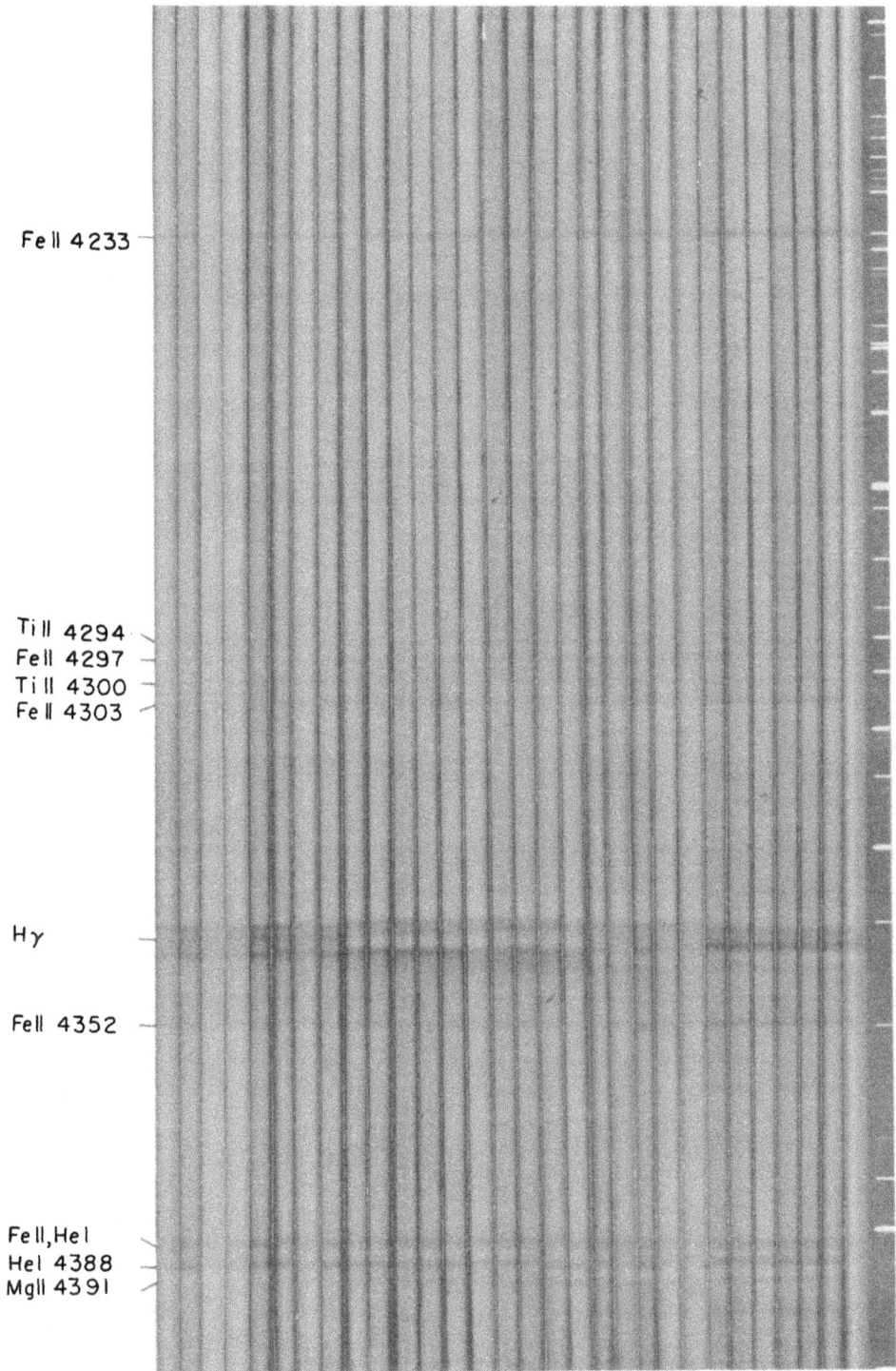

Fe II 4233

Ti II 4294
Fe II 4297
Ti II 4300
Fe II 4303

Hγ

Fe II 4352

Fe II, He I
He I 4388
Mg II 4391

Phase

4.5687
5725
5754
5794
5828
5862
5894
6498
6532
6564
6595
6624
7288
7323
7355
7386
7422
8066
8099
8130
8163
8196
8802
8826
8852
8882
8906
8923
8940
8958

54

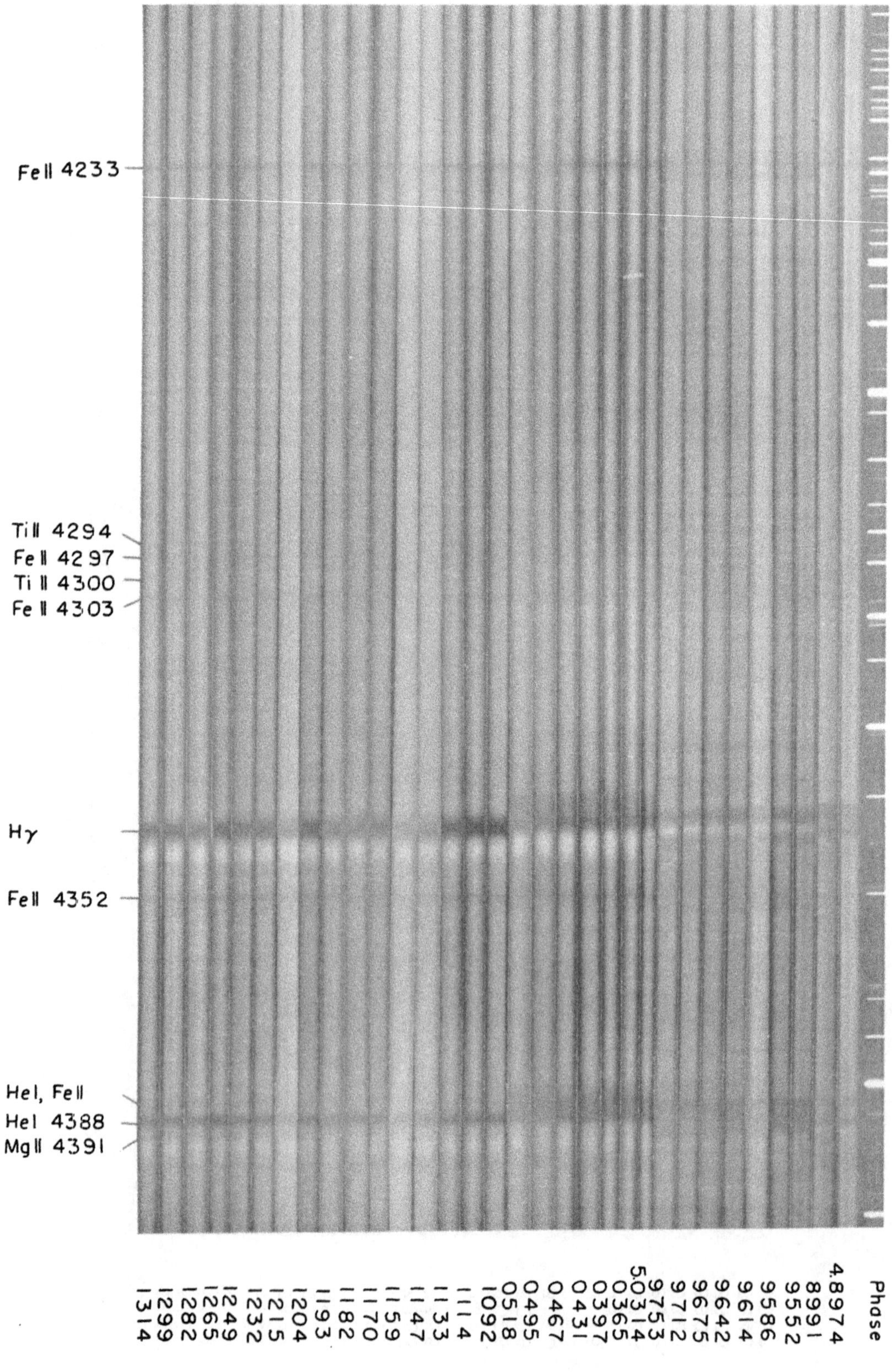

Fe II 4233

Ti II 4294
Fe II 4297
Ti II 4300
Fe II 4303

Hγ

Fe II 4352

He I, Fe II
He I 4388
Mg II 4391

Phase

4.8974
.8991
.9552
.9586
.9614
.9642
.9675
.9712
.9753
5.0314
.0365
.0397
.0431
.0467
.0495
.0518
.1092
.1114
.1133
.1147
.1159
.1170
.1182
.1193
.1204
.1215
.1232
.1249
.1265
.1282
.1299
.1314

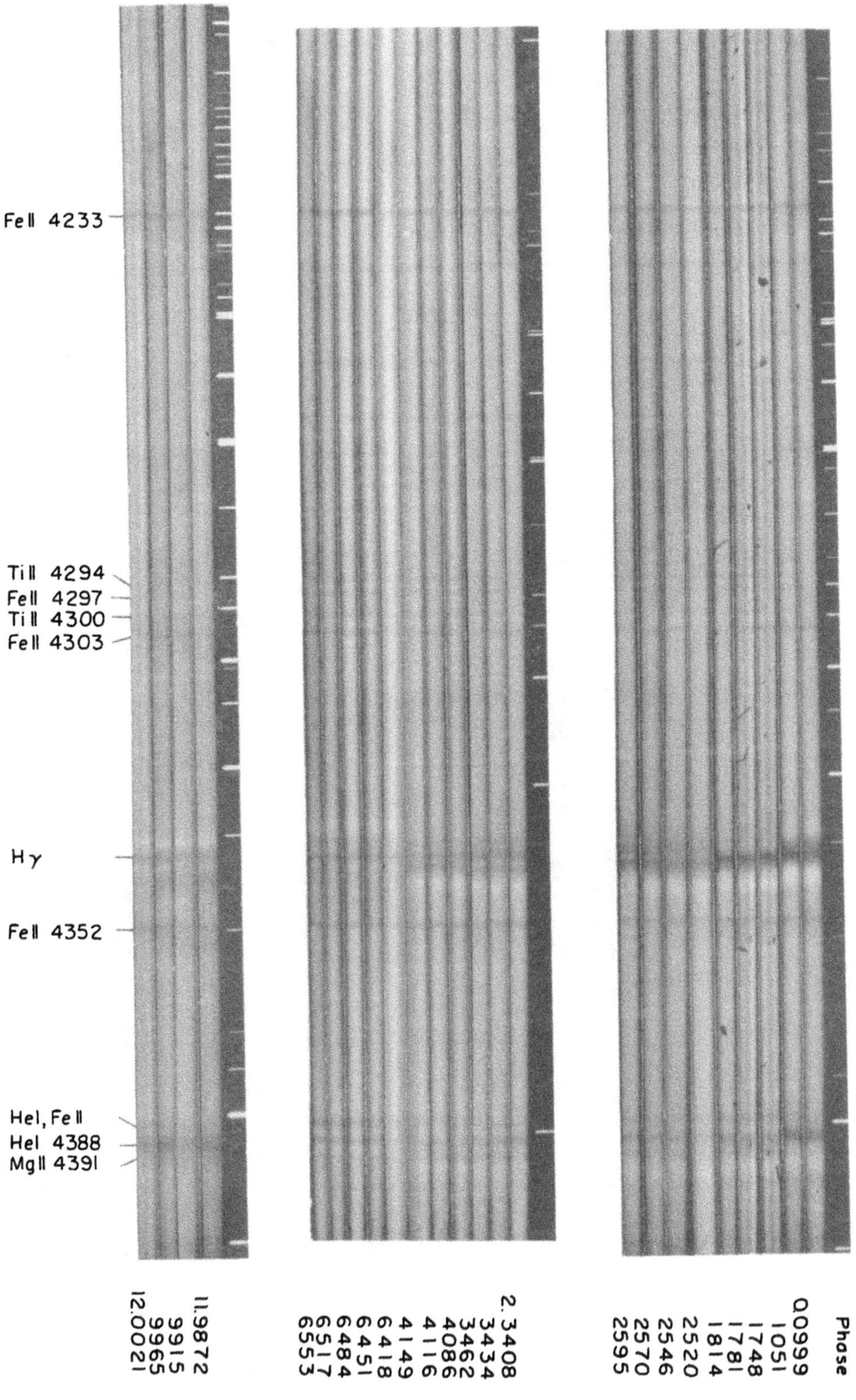

Fell 4233

Till 4294
Fell 4297
Till 4300
Fell 4303

Hγ

Fell 4352

Hel, Fell
Hel 4388
Mgll 4391

11.9872
9915
9965
12.0021

2.3408
3434
3462
4086
4116
4149
6418
6451
6484
6517
6553

0.0999
1051
1748
1781
1814
2520
2546
2570
2595

Phase

53

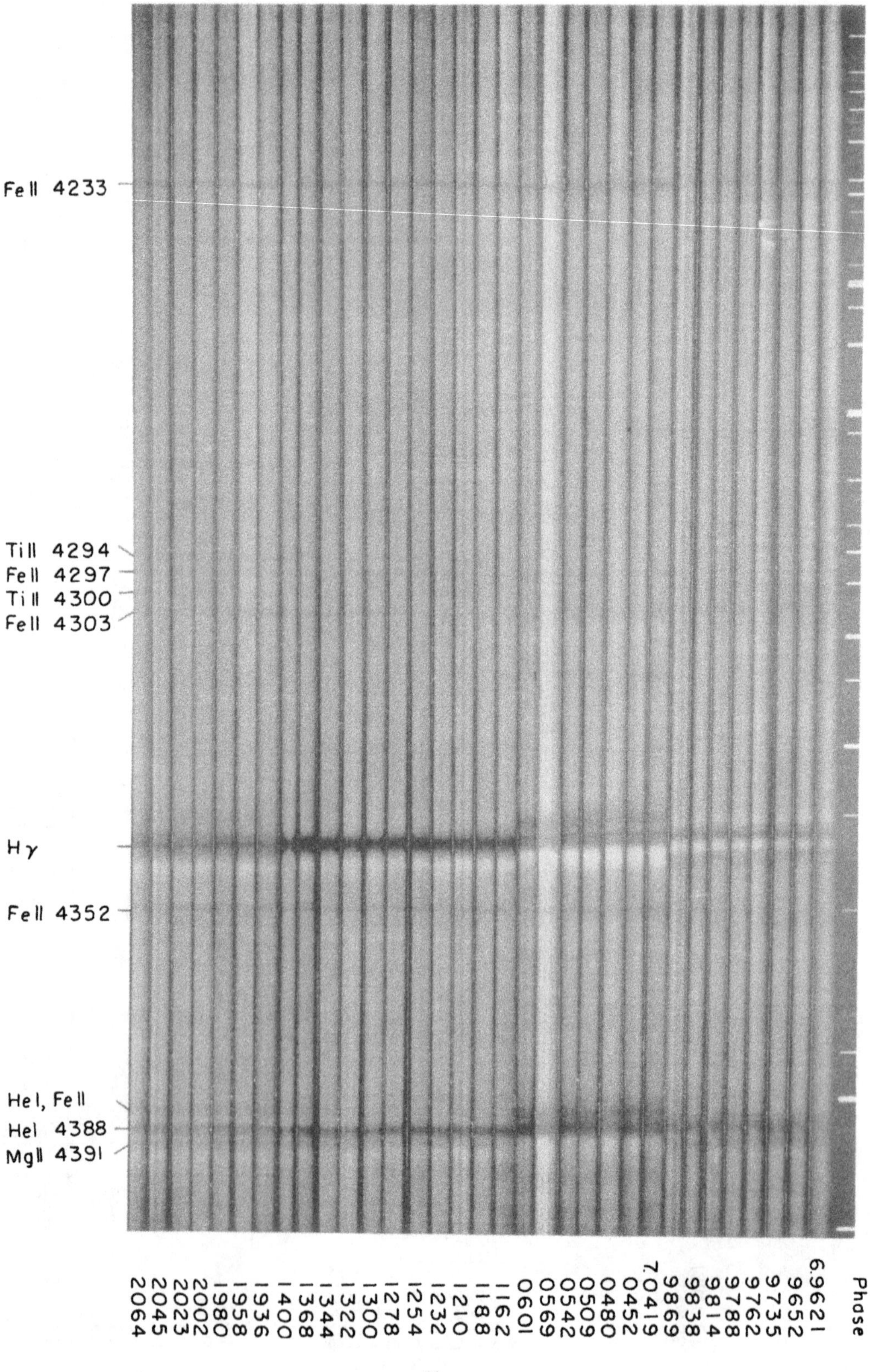

Fe II 4233

Ti II 4294
Fe II 4297
Ti II 4300
Fe II 4303

Hγ

Fe II 4352

He I, Fe II
He I 4388
Mg II 4391

Phase

6.9621
9652
9735
9762
9788
9814
9838
9869
7.0419
0452
0480
0509
0542
0569
0601
1162
1188
1210
1232
1254
1278
1300
1322
1344
1368
1400
1936
1958
1980
2002
2023
2045
2064

56

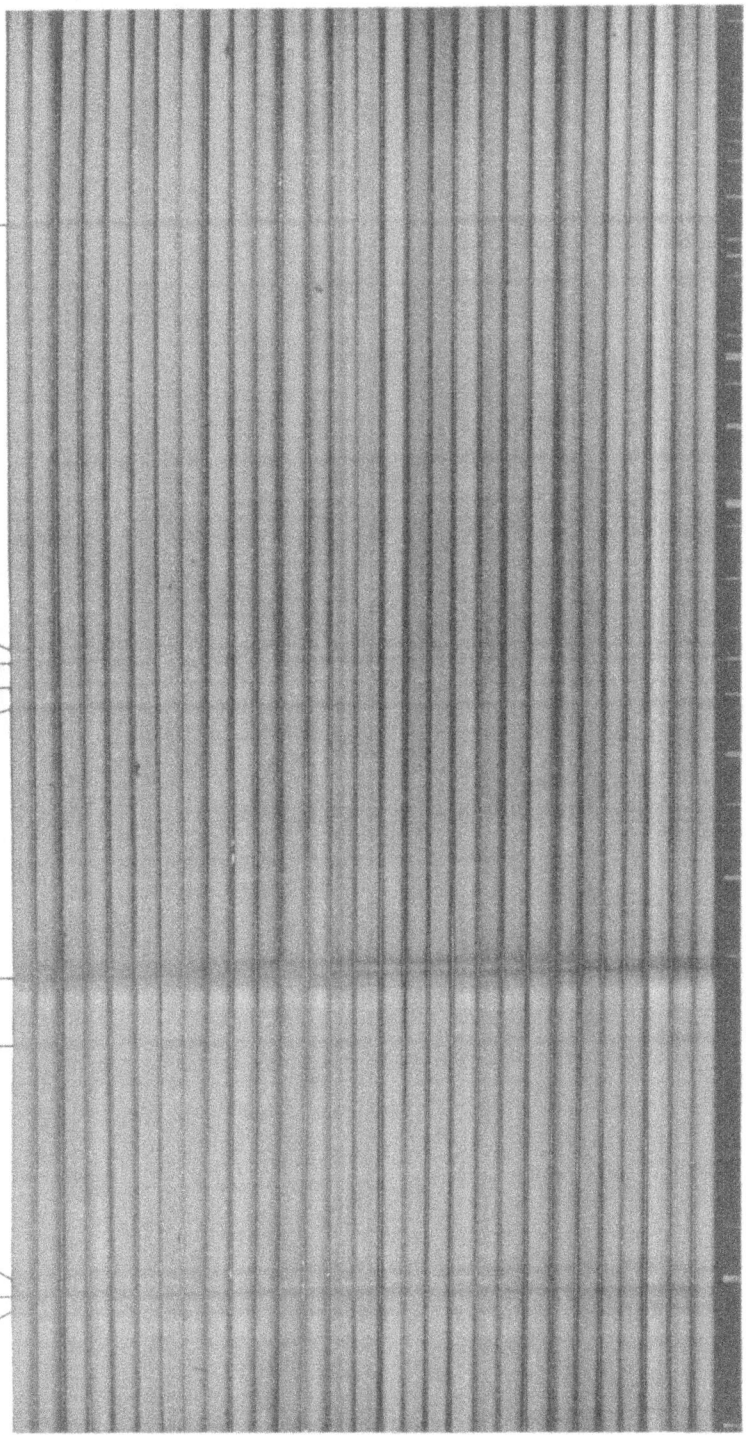

Fe II 4233

Ti II 4294
Fe II 4297
Ti II 4300
Fe II 4303

Hγ

Fe II 4352

He I, Fe II
He I 4388
Mg II 4391

Phase

7.2083
.2101
.2120
.2139
.2161
.2186
.2707
.2728
.2747
.2763
.2779
.2898
.2915
.2934
.2951
.2967
.3485
.3509
.3530
.3547
.3567
.3585
.3599
.3616
.3631
.3646
.3663
.3682
.3699

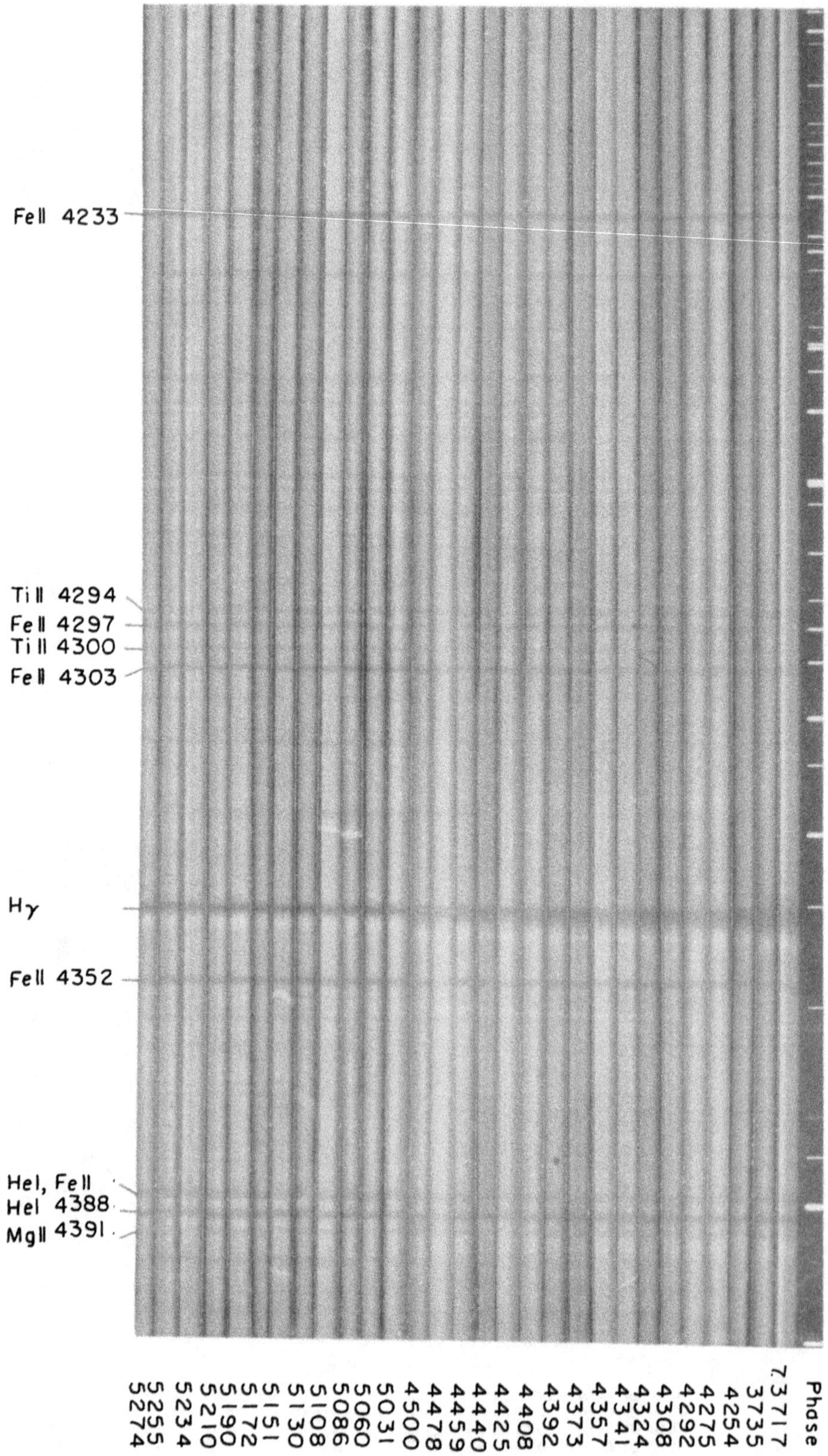

Fe II 4233

Ti II 4294
Fe II 4297
Ti II 4300
Fe II 4303

H γ

Fe II 4352

He I, Fe II
He I 4388
Mg II 4391

Phase
7.3717
3735
4254
4275
4292
4308
4324
4341
4357
4373
4392
4408
4425
4440
4459
4478
4500
5031
5060
5086
5108
5130
5151
5172
5190
5210
5234
5255
5274

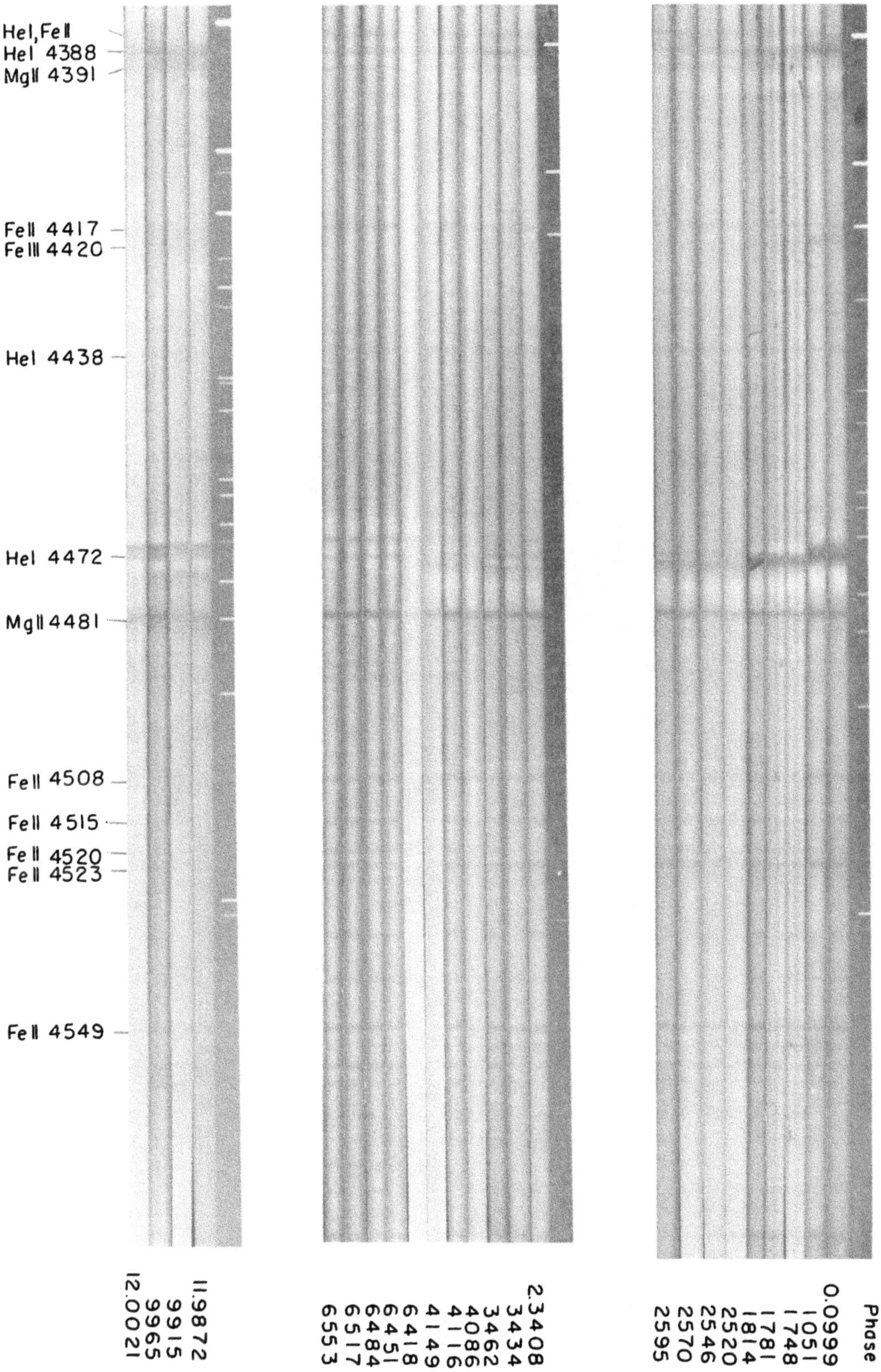

HeI,FeII
HeI 4388
MgII 4391

FeII 4417
FeIII 4420

HeI 4438

HeI 4472

MgII 4481

FeII 4508

FeII 4515

FeII 4520
FeII 4523

FeII 4549

11.9872
9915
9965
12.0021

2.3408
3434
3462
4086
4.116
4.149
6418
6451
6484
6517
6553

0.0999
1051
1748
1781
1814
2520
2546
2570
2595

Phase

59

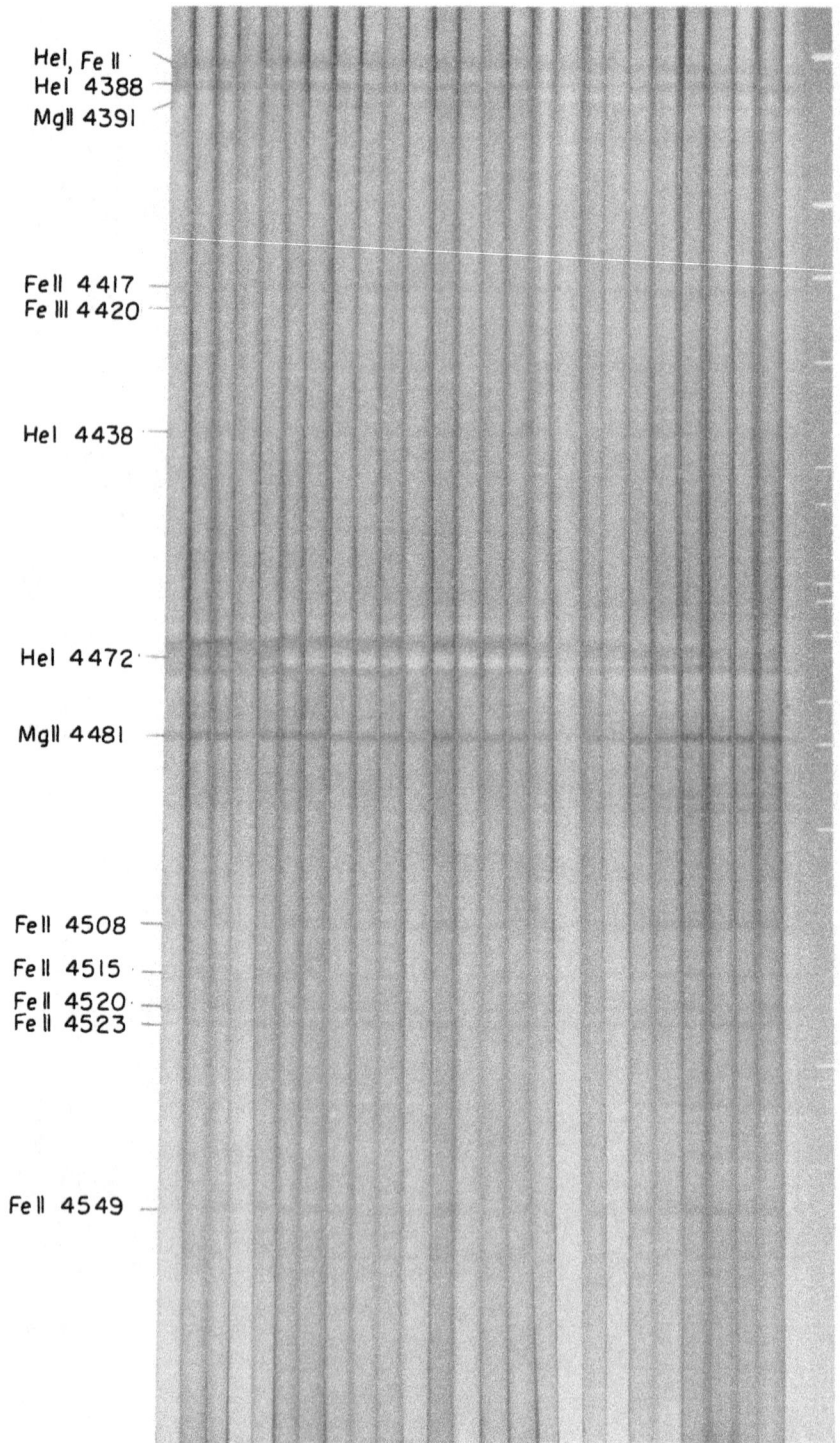

HeI, FeII
HeI 4388
MgII 4391

FeII 4417
FeIII 4420

HeI 4438

HeI 4472

MgII 4481

FeII 4508

FeII 4515

FeII 4520
FeII 4523

FeII 4549

Phase 4.5687
5725
5754
5794
5828
5862
5894
6532
6564
6595
6624
7288
7323
7355
7386
7422
8066
8099
8130
8163
8196
8802
8826
8852
8882
8940

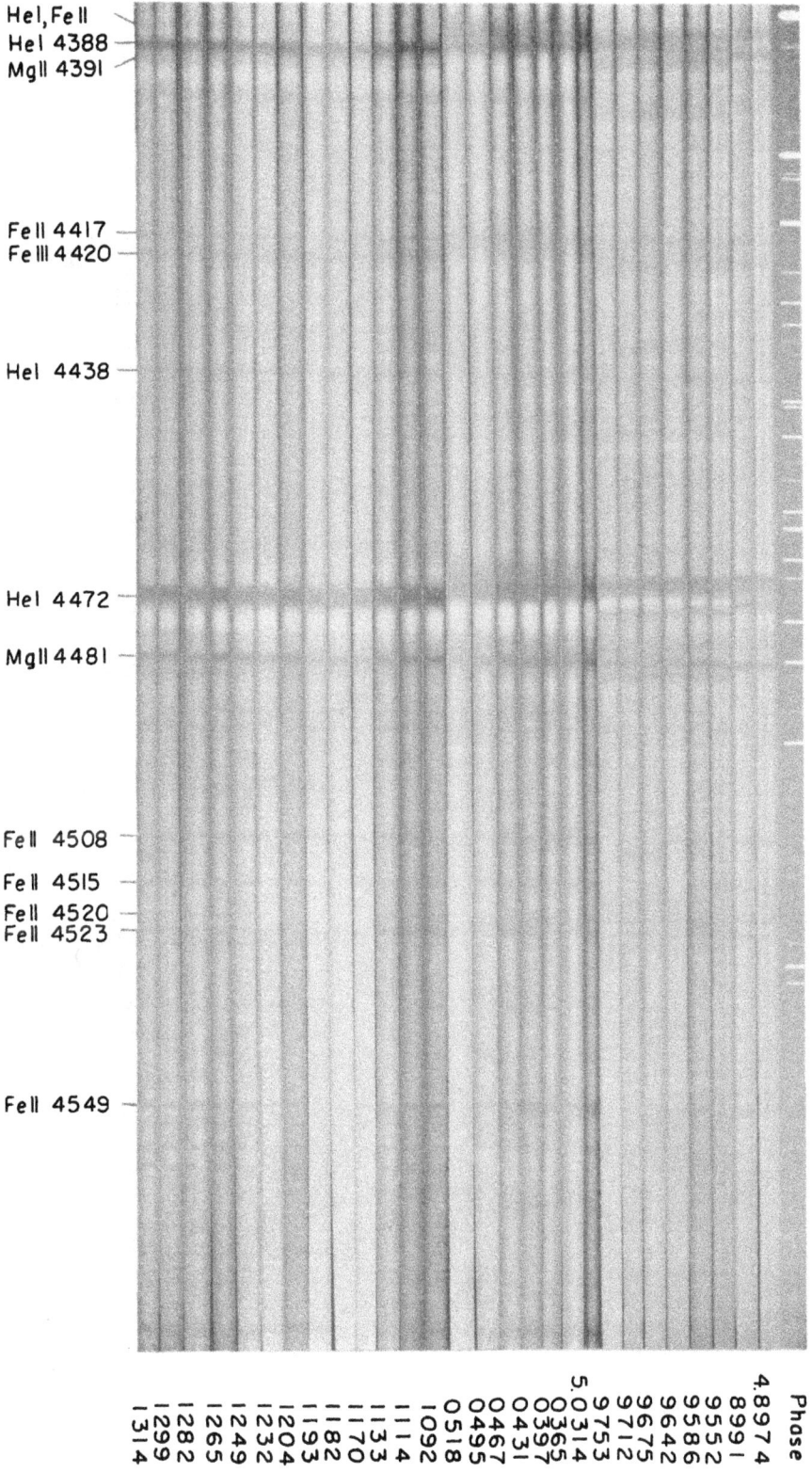

HeI,FeII
HeI 4388
MgII 4391

FeII 4417
FeIII 4420

HeI 4438

HeI 4472

MgII 4481

FeII 4508
FeII 4515
FeII 4520
FeII 4523

FeII 4549

Phase

4.8974
8991
9552
9586
9642
9675
9712
9753
5.0314
0365
0397
0431
0467
0495
0518
1092
1114
1133
1170
1182
1193
1204
1232
1249
1265
1282
1299
1314

HeI,FeII
HeI 4388
MgII 4391

FeII 4417
FeIII 4420

HeI 4438

HeI 4472

MgII 4481

FeII 4508

FeII 4515
FeII 4520
FeII 4523

FeII 4549

Phase

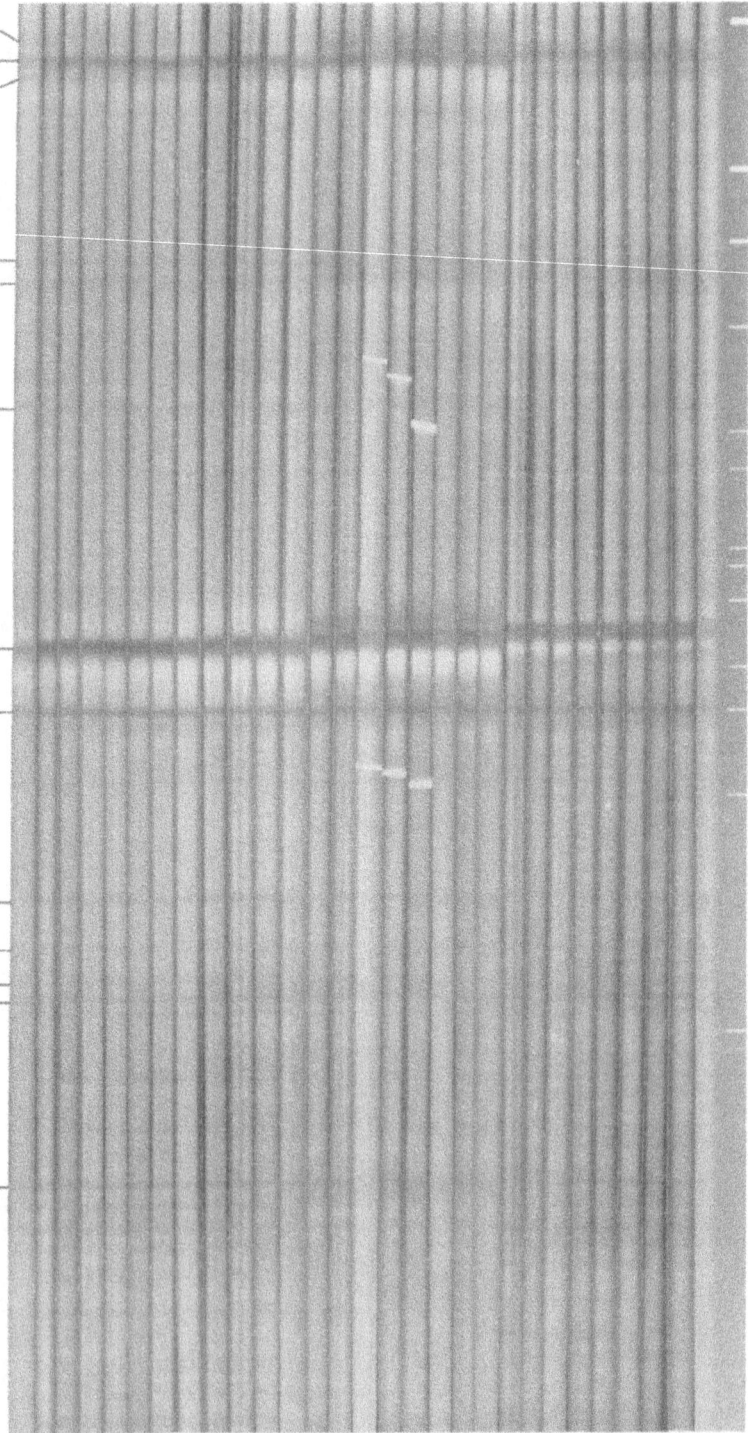

6.9621
9652
9693
9735
9762
9788
9814
9838
9869
9898
.0419
0452
0480
0509
0542
0569
0601
0639
1162
1188
1210
1232
1254
1278
1300
1322
1344
1368
1400
1936

7.0419

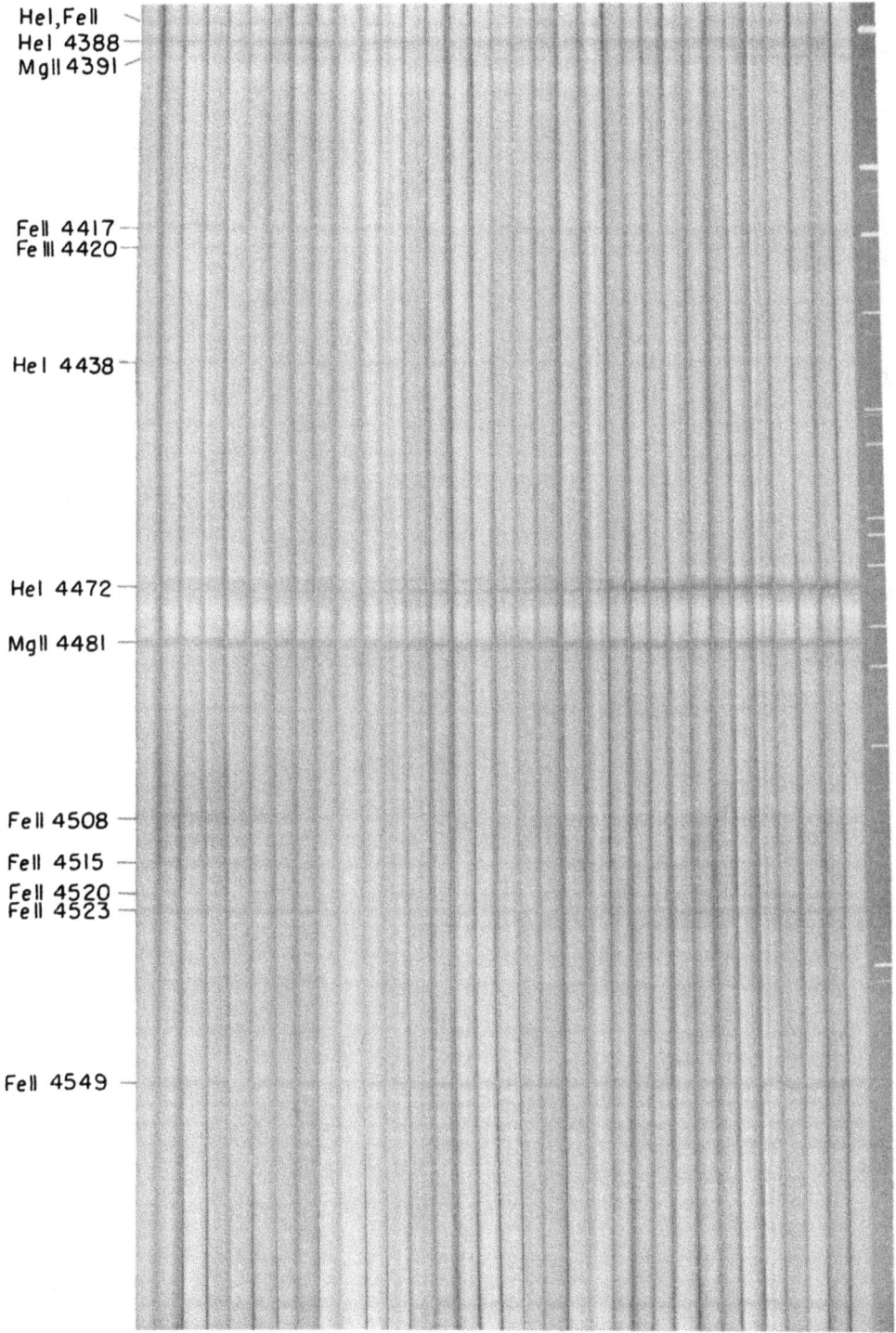

HeI,FeII
HeI 4388
MgII 4391

FeII 4417
Fe III 4420

HeI 4438

HeI 4472

MgII 4481

FeII 4508

FeII 4515

FeII 4520
FeII 4523

FeII 4549

Phase
7.1958
1980
2002
2023
2045
2064
2083
2101
2120
2139
2161
2186
2707
2728
2747
2763
2779
2898
2915
2934
2951
2967
3485
3509
3530
3547
3567
3585
3599
3616
3631
3646
3663

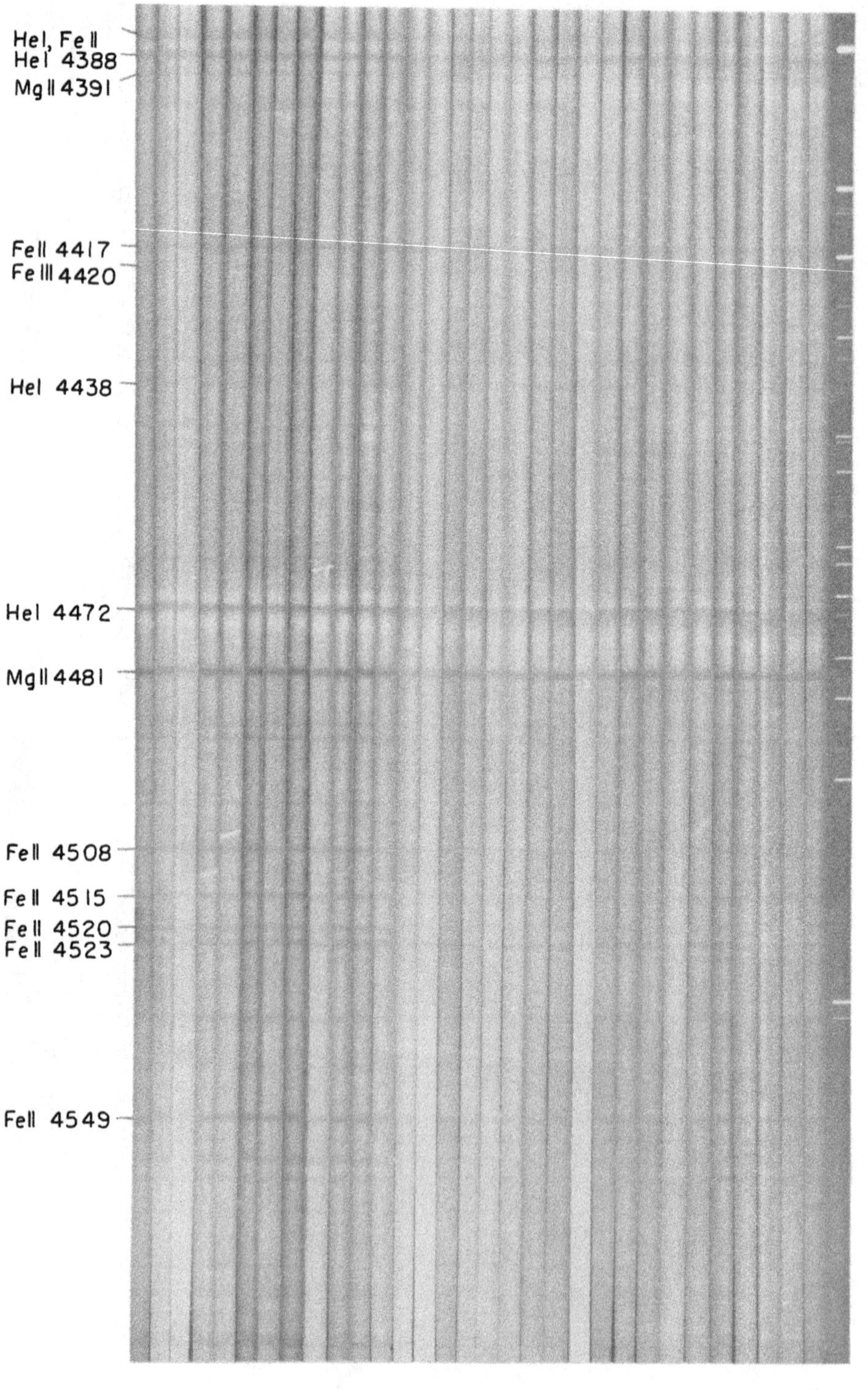

HeI, FeII —
HeI 4388 —
MgII 4391 —

FeII 4417 —
FeIII 4420 —

HeI 4438 —

HeI 4472 —

MgII 4481 —

FeII 4508 —

FeII 4515 —
FeII 4520 —
FeII 4523 —

FeII 4549 —

5274
5255
5234
5210
5190
5172
5151
5130
5108
5086
5060
5031
5000
4478
4459
4440
4425
4408
4392
4373
4357
4341
4324
4308
4292
4275
4254
3735
3717
3699
73682

Phase

www.ingramcontent.com/pod-product-compliance
Lightning Source LLC
Chambersburg PA
CBHW081333190326

41458CB00018B/5988